SOVIET SCIENCE

SOVIET SCIENCE

Zhores A. Medvedev

W · W · NORTON & COMPANY · INC·
NEW YORK

First Edition

Library of Congress Cataloging in Publication Data

Medvedev, Zhores Aleksandrovich.
  Soviet science.

  Includes bibliographical references and index.
    1. Science—Russia—History.   I. Title.
Q127.R9M417 1978   509′.47   78–1628
ISBN 0–393–06435–2

designed by Paula Wiener

1 2 3 4 5 6 7 8 9 0

To the memory of my teachers in science
and friends in life

P. M. ZHUKOVSKY (1879–1975)
V. M. KLECHKOVSKY (1900–1971)
B. L. ASTAUROV (1904–1974)

Contents

Introduction

In this book I have not attempted an exhaustive examination of the entire history of Soviet science—one author could hardly master the diversity of specialized knowledge which would be required for such a task. Nor have I tried to present a condensed account of the main events in the development of Soviet science (with international science as a background for comparison) in the manner of a professional historian. This would be an appropriate task for groups of official Soviet historians of science or the Institute of the History of Science, which is within the framework of the Academy of Sciences of the USSR. Many books have already been written on this subject, both specialized monographs as well as works published to commemorate various important jubilee occasions. One of the last such volumes—the official history of the Academy of Sciences of the USSR—recently appeared to mark the two hundred and fiftieth anniversary of the founding of the Russian Academy (1).

A number of books dealing with the history of Soviet science in general or with specific aspects of it have been published by foreign authors (2–7); unique aspects of the whole phenomenon have obviously intrigued many scholars.

I decided to take up this subject once again out of a feeling that my own approach would be from a specific and rather different point of view. I am not an official Soviet historian who might attempt to magnify various achievements while avoiding real analysis and concealing any failures wherever possible. Nor am I a foreign historian who certainly would have the freedom to deal with the subject in a genuine way through a study of documents and memoirs, political decisions and programs, but who would have no personal or emotional involvement.

Either type of book could be of interest to specialized readers, but it would not appeal to a general audience, even if those who are interested in the problems of science as such were included.

The history of Soviet science covers a period of time which amounts to less than a human life span, and there are a number of senior academicians who distinguished themselves soon after the Revolution and in the 1920s who still work as directors of important research institutes and centers (in the USSR there is no compulsory retirement age for academicians). V. A. Engelgardt (eighty-six years old), at present the director of the Institute of Molecular Biology, did his most important work about fifty or sixty years ago; N. N. Semenov (seventy-nine), director of the Physical Chemistry Institute, and P. L. Kapitsa (seventy-eight), director of the Institute of Problems of Physics, also started their research work in 1923–1924. A. I. Oparin (eighty-seven), who is still director of the Institute of Biochemistry in Moscow, became well known after his theory of the origin of life was published in 1924. Many thousands of Soviet scientists of the next generation, now in their fifties, lived through Stalin's terror and have not forgotten that experience. Westerners and especially Western scientists who read this book might find it interesting to compare the events of their own lives with those of their Soviet colleagues.

Young readers might find it worthwhile to learn about the fate of the intellectual community in a country which has played such an important historical role for the past sixty years.

As a former Soviet scientist who published his first research paper in the *Proceedings of the Soviet Academy* in 1948 (a most dramatic and tragic period for the whole Soviet research establishment), and who worked within the USSR until 1973 (the peak year of *détente* optimism with regard to international cooperation in science and technology), I thought I might be able to tell the story of Soviet science in straightforward language to be understood by a layman, without too many technical details. I have now had the experience of five years of research work in Britain and have traveled extensively, visiting many universities and research centers in the United States and Europe. There have been many conversations about Soviet science, conversations which I hope have helped me to find an approach to the subject which is appropriate for a Western reader. It will, of necessity, be a condensed account; indeed, the length of the text has been intentionally restricted in the hope that it will be read rather than simply stored on a shelf as a possible source of reference.

For this is not merely a book. It is also an appeal, as in the case of the first *samizdat* version of my work on Lysenko in 1962, while he was still in power. When that book was published in the West several years later with additional chapters on the fall of Lysenko (8), it appeared to be a history of the whole Lysenko affair. However, it was not originally conceived as a history. Rather, it was a desperate appeal to attract public attention to the plight of Soviet biology and in this way help to make the fall of Lysenko a real possibility. My second book devoted to Soviet science (published in the West in 1971 with a rather trivial title invented by the publisher, *The Medvedev Papers,* and a more relevant subtitle, "The Plight of Soviet Science Today" [9]) was also an appeal and not simply an account of the politically motivated isolation of the Soviet scientific community. This isolation is now less evident and the policy of *détente* has certainly had a positive influence. But Soviet science is still troubled by a number of old as well as new problems and controversies, and they should be discussed openly. It is easier to understand these issues in their historical context, since it is a rather unusual, often tragic, history, which still has an impact on the current situation.

Of course, the misuse of science and technology is by no means a unique phenomenon specific to one country alone. Many other countries with different social systems have, in a variety of ways, manipulated scientific progress for political ends. I should like to make it clear for the readers of this book that further measures must be taken in all countries to protect science from possible misuse. Truly, science can be the source of good or evil in human life, and all scientists share a responsibility for our future.

SOVIET SCIENCE

Chapter 1

The Socialist Revolution
and Russian Science (1917–1921)

Almost as soon as the new Soviet system came into being after the Bolshevik Revolution in October, 1917, the role of science and the position of scientists in society became a source of controversy. Before the Revolution, Russian science had largely been an elite establishment within the state structure. All types of research units in the natural and technical sciences, in universities, institutes, commissions, and so on, were supported by the state and formed a hierarchical pyramid with the Imperial Russian Academy of Sciences in the most privileged supreme position.

In this poor, predominantly peasant country, members of scientific institutions, professors in the universities, in fact all those with scientific degrees at a postgraduate level, could count on having a very high place in the social structure. The average income of a professor or member of the Academy was about twenty or thirty times higher than that of an industrial worker, and very few among the scientific community belonged to or sympathized with the aims of the radical revolutionary socialist parties.

On the eve of the Revolution, many Russian scientists were strongly opposed to the absolute monarchy in its existing form, but when they participated in political life, most of them associated themselves with the more moderate Constitutional Democrats (Kadets) or other groups in favor of parliamentary democracy, although some did belong to the moderate socialist party, the Mensheviks. Thus the scientific community welcomed the collapse of the monarchy in February, 1917, and the development of a democratic system (largely an imitation of the British example). The Russian Academy (now without the "imperial" in its title) as well as other scientific and educational networks actively cooperated with the provisional government in its attempt to put an end to the anarchy and disorder which prevailed after the collapse of the imperial government. However, in 1917 the main problem for Russia was the war which had already been going on three long years. A large part of the country was occupied by the Germany army. Russia was losing the war, the army was retreating, and the change of government in this most disastrous situation could hardly increase the strength of the Russian troops or make them better able to resist the German advance.

The provisional government, notwithstanding its many positive democratic reforms, pursued a policy of continuing the war "to a victorious conclusion" as a result of Allied pressure; it was this suicide course which made the new, more radical revolution inevitable. The unlimited freedoms of the young democratic state were powerless to prevent eventual military defeat. However, even under strong pressure from the right (in favor of military dictatorship) and from the left (calling for an end to the war and all power to the Soviets), the provisional government was able to institute a number of measures during its very short time in office which supported and strengthened the scientific community. Between March and October of 1917 several new research institutes were established, most of them devoted either to the study of national mineral resources or to the development of those branches of science and technology oriented toward military needs.

The Bolshevik Revolution in October of the same year was not welcomed by the majority of the intellectual and scientific elite. At the same time, the Council of People's Commissars (as the new government was called) and the leadership of the Bolshevik Party were rather suspicious of and hostile toward "bourgeois" scientists and experts. However, those few prominent scientists (such as K. A. Timiriazev, a plant physiologist, or V. R. Williams, a soil scientist) who heartily welcomed the Bolsheviks and their revolutionary ideas received strong support and publicity.

As soon as the government began to carry out its program of economic and political reorganization, it was inevitable that there would be a conflict with the privileged scientific elite. The first three to four months of Soviet power had been rather peaceful, with the Civil War coming only later. The initial measures of the government, its decrees on the end of the war and peace negotiations, on the confiscation of landlord estates, and land reform in favor of the peasants, as well as many other actions, created strong support among workers, peasants, and soldiers. However, various economic measures such as the nationalization of all banks and industry and the amateurish attempt to entirely liquidate the monetary system (to be replaced by a direct exchange of agricultural and industrial products) created an extremely difficult situation. There was a reduction in the supply of agricultural produce to urban areas because farmers' markets in the towns were closed down. In many industrial centers, and in Petrograd and Moscow as well, there began to be a rationing of food supplies. The Extraordinary Commission (Cheka) was created in order to prevent acts of sabotage, strikes, speculation, and other "counterrevolutionary" acts and was given the powers of summary trial and execution. Government by administrative command gradually came to be backed by "red terror."

The peace treaty signed with Germany at Brest was humiliating even for a defeated country—a large part of Russian territory was ceded to the German empire. The majority of intellectuals as well as officers in the army had been opposed to this kind of peace treaty; in the countryside, where there had been little serious violence during two major revolutions, the situation began to deteriorate rapidly. Small military uprisings in different parts of the country, together with strong peasant resistance to the confiscation of agricultural products, soon developed into full-scale civil war. Although receiving little attention in the midst of all these major confrontations, the scientific community was in a deeply divided state. The larger part of senior research and academic personnel backed the anti-Bolshevik forces, and during the first waves of "red terror," "professors" and "academicians" were almost automatically considered to be enemies of Soviet power. A large number of scientists and technical experts were harassed, arrested, sentenced, and even executed during the beginning of the Civil War in 1918–1919. Professor Nikolai Koltsov, a famous cytologist and biologist, was, in 1918, under arrest and sentenced to death in Moscow. The only reason for this was the fact that he had been a Kadet. His life was saved, however, because his close friend, the famous writer Maxim Gorky, appealed on Koltsov's behalf directly to Lenin. (Gorky organized a special committee to help prominent scientists survive the two major perils of the time, terror and famine.)

The anarchy of civil war and the instability of Soviet power during 1919–1921 forced many prominent Russian scientists and intellectuals who had not been on the "red" side to emigrate or simply flee abroad. Some scientists and intellectuals were expelled from the country (at that time deportation was considered to be a punishment almost equal to death). Within a short time the young country lost many famous figures. Hundreds of names could be mentioned, but it is enough to indicate a few of the more famous ones to show the scale of this intellectual exodus. Among those who left Russia was Igor Sikorsky, prominent aircraft engineer, who was the first to design the multimotor plane in 1913. He crossed the border in 1919, settled in the United States, and soon became a leading figure in aviation technology as designer of the helicopter. Professor V. Korenchevsky, a biologist at Petrograd University, emigrated to Britain, where he soon became the world's leading specialist on the problems of aging. He organized many gerontological societies as well as the International Association of Gerontology. Because of his many important works in this field, he became known as the "father of gerontology." G. B. Kistiakovsky, who left Petrograd as a young chemist, later distinguished himself in the United States and became a vice-president of the National Academy of Sciences and a scientific adviser to President Eisenhower. The leading Western sociologist Pitirim Sorokin, the economist and future Nobel Prize winner W. Leontiev, and many others all left Russia during the turbulent years of the Civil War. Russian art, literature, and music were also deprived of their most famous talents during these years. Ivan Bunin, the future Nobel laureate for literature; Igor Stravinsky, Sergei Prokofiev, Sergei Rachmaninoff—three great composers and conductors; Anna Pavlova, George Balanchine, Feodor Chaliapin, Marc Chagall, Vassily Kandinsky, Vladimir Nabokov—these were among the outstanding figures who left Russia and later made their very special mark on the development of international science and culture. Even the famous "proletarian" writer, Maxim Gorky, left the country immediately after the Civil War and later wrote many bitter letters to Lenin about abuses of terror (these letters have never been published and are in the secret holdings of the Marx-Lenin Institute in Moscow).

This is just to mention a few; in fact many thousands left the country or were forced to leave by the end of the Civil War. Those who went abroad were at least able to continue their work and remain alive; when the New Economic Policy (NEP) made life in Russia more tolerable, some of them returned (Gorky, Alexei Tolstoy, later Prokofiev, and others—on the whole, writers and artists rather than scientists). But many intellectuals who supported the "whites" during the Civil War were executed or killed in the course of bloody battles. One of the most tragic events during the wave of Civil War "red" terror was the execution of the great Russian poet Nikolai Gumilev, sentenced to death by a firing squad in Petrograd in 1921. Gumilev had been accused of participating in a counterrevolutionary group (the accusation has now been proved to be false, the "group" never in fact having existed at all), and was arrested together with more than sixty other intellectuals, scientists, and some officers and soldiers in August of 1921. Within two weeks all of them were secretly tried and sentenced to death. The "group" included several professors, among them N. I. Lasarevsky, M. N. Tichvinsky, and V. M. Koslovsky.*

The privileged position of scientists and experts in imperial Russia and their certainty of belonging to the "middle class" (in the Western sense of this term) made it natural for them to oppose the policy of War Communism which was dominant during 1918–1921. But this policy was not a result of the Civil War. Quite the contrary. In many cases it was in fact the cause of uprisings in rural areas of Russia and served to prolong the Civil War because not only the middle class but also the majority of the Russian peasantry were against this primitive "communist" system.

The professional scientific community of Russia before the Revolution consisted of a little more than eleven thousand people, and most of them lived in St. Petersburg (renamed Petrograd in 1914) or Moscow. However, their knowledge and expertise was of a very high standard, and Russian science enjoyed an excellent reputation in Europe. But in tsarist Russia education, and particularly higher education, was largely the privilege of certain social groups, and one of the slogans of the new Bolshevik administration was "knowledge and education for the masses." Thus, from the very beginning of Soviet power, members of the

---

* The official report of the "liquidation" of this "group," with a list of those executed, was published in *Petrogradskaya Pravda*, No. 181, Sept. 1, 1921.

prerevolutionary scientific elite were persecuted in spite of the fact that this could not fail to undermine the large-scale program of expansion in education, science, and technology.

The call to develop a "new" science and also to eliminate the "bourgeois" intelligentsia clearly involved a conflict of goals which would have to be resolved. The real danger of the "brain drain" as a result of civil war and emigration became all too apparent as soon as the Bolsheviks began to take measures to restore industry for military purposes in order to fight the dangerously prolonged Civil War. Any large-scale war needs technological support, and civil wars are no exception. The change of attitude toward military and scientific experts became evident from the beginning of 1919 and particularly at the Eighth Congress of the Bolshevik Party. Lenin introduced the resolution which recommended that the ideological approach be dropped when dealing with technicians and scientists. The resolution was approved and became the official line. It declared:

> The problem of industrial and economic development demands the immediate and widespread use of experts in science and technology whom we have inherited from capitalism, in spite of the fact that they inevitably are impregnated with bourgeois ideas and customs [10].

It was hoped that this new approach would serve to accelerate processes of education and research, training a new "revolutionary" generation of scientists and technical experts who would later be able to replace "bourgeois" scientists, engineers, and intellectuals.

The privileged economic position of scientific and technical experts had been undermined by the rapid inflation which made new Soviet money worthless soon after the Revolution. However, from 1919 professional personnel employed in established and also newly created educational and scientific institutions were given special rations and financial support—a concession the Soviet government understood to be inevitable in order to discourage emigration but also simply to make life, and thereby work, possible.

A historian of science might well be astonished by the number of new educational and research institutions created during the most dramatic period of the Civil War when the very existence of the Soviet system was seriously at stake. The battles of two successive wars had wrecked the economy of the country, and the task of restoration seemed almost impossible. However, it was assumed that science and technology would be the major factors in postwar reconstruction and "electrification" of the country, and this assumption relieved the situation of those scientists who chose to cooperate with the new administration. They received enormous support, taking into consideration the limited means of an impoverished country. There was, however, no shortage of space for new research centers, schools, and colleges. The abandoned palaces, houses, and estates of the Russian aristocracy and wealthy elite who had fled the country or had simply been killed were not always appropriate for conversion into apartment blocks for workers, and a number of them in Petrograd, Moscow, and other cities had been transformed into schools, institutes, research units, and laboratories. Moscow and Petrograd universities, the Academy of Sciences, and other already functioning scientific centers acquired many new and rather elegant buildings and palaces.

The way in which the young professor Nikolai Vavilov (later to become a famous scientist) was able to establish the new institute of plant breeding in Petrograd in 1920 was a typical story of the times.

Vavilov worked in the provincial Saratov University but was already well known for his abilities in botany and genetics. In 1919 he was invited to Petrograd to head the Laboratory of Applied Botany. The former head of the laboratory was the famous botanist and academician Robert Regel, who died from typhus—the main health hazard during the war. Nikolai Vavilov had been a personal friend of Regel and was extremely bitter about his fate. He wrote and published an obituary which expressed his grief and also his alarm about the fate of Russian science: "With every day that passes the ranks of Russian scientists grow thinner and thinner and the fate of Russian science lies in the balance. Replacements are many, but few of them are real" (11).

Although he lived in Saratov, Vavilov had been a deputy of Robert Regel, and after the tragic death of his teacher he left Saratov and moved to Petrograd, then an almost deserted city with only one-third of its prewar population. The central government had already been transferred to Moscow. St. Petersburg (Petrograd) had been the capital of Russia only since the beginning of the eighteenth century, when Peter the Great began to build the city as a "window to Europe." The 1918 move to Moscow was largely motivated by strategic considerations; Petrograd had become too vulnerable after Finland and the Baltic provinces of Russia became independent and could serve as bases for anti-Bolshevik forces.

The small center of applied botany was situated in a half-destroyed house which no longer had water or heating. However, many large buildings were empty and available to the energetic new director. His choice was approved by the authorities, and his Laboratory of Applied Botany was installed in a large palace near the center of the city—the former office of the Ministry of Agriculture. It was a beautiful, enormous three-story house with more than a hundred large rooms, a library, offices, and lecture halls. Very soon Nikolai Vavilov organized the new Institute of Experimental Agronomy in Petrograd and Moscow; in 1929 it was to become the Academy of Agricultural Sciences with Vavilov as its first president.

Vavilov's experience was characteristic of the time. In spite of an extremely difficult situation, the Revolution created many possibilities for enthusiastic scientists. The new Research Institute of Physico-Chemical Analysis was established in 1918, and the Research Institute of Platinum and Other Precious Metals was founded at the same time. In 1919 several new institutes were created within the Academy of Sciences, including the Institute of Optics and the Institute of Radium. The Research Institute of Roentgenology (which later became the center of atomic research) was established in 1920, the Research Institute of Physics and Mathematics in 1921, and so on. Similar developments could be observed in education, in experimental stations, and in the formation of special commissions for various research and development programs. The new Ukrainian Academy of Sciences was also established in 1919.

Many research institutes were set up in the applied sciences. In 1918–1919, in Moscow, the Electro-Technical Institute, the Academy of Mining and Engineering (Gornaya Academiya), the Central Aerohydrodynamic Institute (for research in aviation technology), and others were founded. The First National Congress of Physics took place in Petrograd in 1919 and received a great deal of publicity in the Soviet press.

In Petrograd, in 1920, several academicians supported by a group of young scientists organized a permanent "Atomic Commission." This commission tried to coordinate research on the structure of atoms. Young scientists who started to work with this commission and the leading Russian physicists associated with it (Professor A. F. Ioffe, D. S. Rozhdestvensky, and others) later played an important role in the Soviet military atomic program. Igor Kurchatov, the head of the program and the "father" of the Soviet atomic bomb, joined this group in 1925.

The creation of large research institutes supported by the Soviet government was very much an innovation in the organizational structure of international science. In most other countries research was carried out under the aegis of universities and colleges and was not coordinated by any government body.

Chapter 2

The Golden Years of
Soviet Science (1922 – 1928)

The year 1921 not only marked the victorious end of the Civil War, it was also the year of economic crisis, of the complete failure of the system of "war communism." The disrupted country was hardly able to survive the economic disaster created by hyper-inflation and the collapse of normal relations between industrial and agricultural producers. War Communism had proved to be a mistake, and to restore normality and repair the damage everywhere it was necessary to return to a more traditional economic policy. But the New Economic Policy, as it was called, was actually a backward step which reintroduced state and private trade, legalized private enterprise production, and accepted a capitalist economy as competitive and complementary to the socialist system. This major change meant an inevitable liberalization of the political climate, while financial reform put an end to inflation; the introduction of gold backing for bank notes was one of many measures which quickly stimulated the country's economic development.

This new situation, coupled with strong state support for scientific, technological, and educational growth, created an extremely favorable climate for the development of Soviet science.

In January, 1921, Lenin discussed some scientific development problems with leading scientists of the Academy of Sciences and of the Navy Medical Academy. Their discussion covered the restoration of international cooperation, the exchange of scientific literature, financial and material support for new institutes and laboratories, a step-up in scientific training, and many other measures. A few days before this meeting a special decree of the Soviet government provided new facilities and unlimited support for the physiological research of the Nobel Prize winner and academician Ivan Pavlov, who was working in Petrograd. A special Foreign Science and Technology Study Bureau had been set up to evaluate the gap created by years of isolation during the Civil War.

If one compares the rate of scientific and technological development in the Soviet Union between 1922 and 1928 with that in a similar period before the Revolution, one finds a tremendous acceleration in research and education programs. Some historians have called it the "Scientific-Cultural Revolution." Ideological orientation and pressure were very weak and were felt mainly in the humanities. Most research institutes, laboratories, departments, university and technical college faculties, learned societies, and technical bureaus were headed, as a rule, by the representatives of the old scientific elite, the "bourgeois" specialists. Objective historians in the West acknowledge that before 1929 ideological pressure "was exerted upon scientists only gradually and indirectly, largely through the slowly growing number of Bolshevik-minded graduate students" (5).

In the humanitarian scientific fields the pressure of ideological control was much more serious. However, it was felt not through the networks of the Academy of Sciences or of Moscow and Petrograd universities, but by means of the creation of special Marxist educational and research centers, such as the Communist Academy of Social Sciences and the Institute of Red Professorship, plus a few so-called "Communist universities," which mainly prepared students for administrative, party, trade-union, and Comsomol positions.

The Academy of Sciences of the USSR, whose main base was still in Petrograd (renamed Leningrad after 1924), retained considerable prestige but became less influential as a research establishment. Several new academies were created, including the All-Union Academy of Agricultural Sciences, the Academy of Medical Sciences, and the Ukrainian Academy of Sciences, and these, together with research centers for applied sciences, which often had a very wide network of experimental stations and design bureaus, became a dominant force in technical and scientific development. Academic and scholarly titles and degrees, such as "academician," "professor," and "doctor," had been devalued after the Revolution, and the new generation in the research community did not help to restore their former prestige. Income and influence in the scientific community during the 1920s were not directly related to degrees and titles (as was the case after the Second World War), and the whole "standard of living" question had a very different meaning. Most people tried to be very modest in their demands. The phrase "barefoot scientists" was coined at this time and really meant something.

In the Imperial Russian Academy of Sciences all academicians had received regular payments from the state merely for membership. These payments were customary in royal academies in prewar continental Europe. In Russia in the early eighteenth century the "academician's pay" was also necessary because most early members had been invited from abroad. However, when the new "charter" of the Academy was introduced a century later, in 1836, the salary of a full academician was set extremely high, at 5,000 rubles per year—an income which put academicians on a par with the Russian aristocracy.

All these privileges were eliminated after the Revolution. When new statuses for the Academy were introduced in 1927, the payments for membership were not mentioned. The membership salary was not reintroduced until later, but it was never again regulated by the charters of the Academy, though these were rewritten many times (in 1930, 1935, 1953, and 1963) (12).

After 1920, first the Academy of Sciences, and later some other research centers, took steps to establish direct links with foreign research centers. Although international cooperation was at first rather modest, it was nevertheless extremely important for the development of Soviet science. In the period 1920–1928 visits abroad (like the right of emigration) were not yet restricted by numerous political and bureaucratic fences. The shortage of foreign currency was the main factor limiting the opportunities for official foreign travel. In 1920 the Academy of Sciences sent only ten persons abroad for research and education. In 1922 the figure had increased to seventeen, in 1924 to twenty-five, and in 1926 to forty-four (1).

At the government level, science at this time was directed by Glavnauka, a special department of the People's Commissariat of Education headed by A. V. Lunacharsky, a liberal and well-educated man among Bolshevik leaders. Glavnauka was also responsible for Soviet scientists' foreign travel and contacts. In 1924 the All-Union Society for Cultural Connections with Foreign Countries (VOKS) was established and became responsible for invitations to scientists abroad. The All-Union Council of Industry and Agriculture (Vsesoyuzny Soviet Narodnogo Khoziaistva)—the prototype of the future State Planning Commission—also created a scientific-technical department which became responsible for imports of modern scientific-technical equipment and technical literature. This scientific-technical department (NTO) set up a permanent office in Berlin known as the Foreign Science and Technology Bureau (BINT). BINT sometimes received direct messages from Lenin. Through this bureau Soviet libraries and institutes began to receive about eighty foreign scientific journals, as well as some advanced optical and other research equipment from the Zeiss and Simens companies.

The increasing productivity of Soviet science was reflected in the rapid growth of the number of scientific journals, papers, and books. By 1922 the number of research publications was already four times greater than in 1921, and by 1923 eight times greater (13).

At this stage there was still nothing to suggest the future long isolation of Soviet science. All the tendencies were against it. A frequent traveler in Europe in early 1920 was the academician A. F. Ioffe, who was a founder of the Soviet School of Atomic Physicists. In 1924 he published an article in *Pravda* entitled "Russian Science Abroad" (14), in which he made clear that "Russian scientific works must be part of the general international stream of science. In this stream Russian science must find an important place and influence the general development of world science."

The New Economic Policy led to the speedy recovery of industrial and agricultural development, and this recovery led to more substantial measures for financial support of scientific and technical research. This—and comparative ideological tolerance—created a unique opportunity for real scientific progress, and in 1922–1928 Soviet science made a very good start indeed. It is difficult to cover all fields of science and technology, but the impressive development of many biological sciences was more or less typical of other fields of research.

N. I. Vavilov formulated his important theory of the evolution of plant species in 1920 and published it abroad in 1922. This theory underlay the creation in the USSR of the world's largest collection of cultivated plants as a general genetic pool for selection and hybridization. The major part of the collection, which played an important role in improving the quality of agricultural plants in the Soviet Union and in introducing many varieties of plants never before cultivated in Russia, was acquired in

1922–1929 as a result of numerous successful expeditions by
Vavilov and his colleagues to distant parts of the world, including
South America, the Middle East, Afghanistan, Turkey, India,
China, and North Africa, to collect every variety of plant cul-
tivated there.

In 1927 an American geneticist, H. J. Muller, working with the
fruit fly (Drosophila), proved that X rays are mutagenic. This discovery became the basis of radiation genetics, and Muller was
honored some twenty years later by the award of a Nobel Prize.
However, the first X ray–induced mutations had already been described by the Soviet biologists G. A. Nadson and G. C. Fillip-
chenko in 1925, after they had irradiated some species of yeasts
and other fungi (15).

A Soviet geneticist, S. S. Chetverikov, was the first to formulate
some laws of genetic polymorphism—a discovery which became
the foundation for population genetics. His classic essay on genetic polymorphism was published in the USSR in 1927, a few
years before the works by S. Wright and J. B. S. Haldane in the
same field. Chetverikov's work was a kind of synthesis of evolutionary Darwinism and Mendelian genetics. Although it was not
immediately recognized (we shall talk later about some political
twists in the fate of Soviet genetics which were responsible for
the interruption of Chetverikov's work), Chetverikov's contribution is now well appreciated, and even in foreign textbooks he is
described as the "founder of experimental population genetics."
In an explanatory note about Chetverikov, I. M. Lerner and W. J.
Libby wrote in their textbook on genetics (16):

> Chetverikov's essay included most of the basic ideas on
> which the current evolutionary theory rests. He recognized
> the existence of the great store of genetic variability supplied
> by mutation in natural populations: he understood clearly
> how particulate Mendelian inheritance makes it possible to
> maintain this variation: he visualized the concepts of allele
> frequency and the gene pool, he realized the roles of such
> evolutionary forces as isolation and drift: he emphasized the

importance of such phenomena as pleotropy of gene action, and of polymorphism in population [p. 242].

Chetverikov laid the foundation for the whole school of Russian general and population genetics; he was not alone, however. Among his students at Moscow University were N. V. Timofeev-Resovsky, B. L. Astaurov, and N. P. Dubinin, all of whom later became world-famous geneticists.

In 1927, G. D. Karpetchenko, a collaborator of N. I. Vavilov, produced the world's first synthetic plant, which was named "Raphanobrassica." Karpetchenko used polyploidization of a hybrid between the radish (Raphanus) and the cabbage (Brassica). This method of polyploidy of hybrids was later used by other scientists.

N. K. Koltzov, who narrowly escaped execution in 1918, organized the Research Institute of Experimental Biology. This institute became the center of cytogenetics, and it was Koltzov also who, in 1927, became the first to develop the revolutionary concept of the gene as a giant protein molecule which can be reproduced by a template mechanism—a concept that linked genetics with biochemistry. The concept was formulated by Koltzov at the first meeting of the All-Union Congress of Zoologists, Anatomists, and Histologists in Leningrad on December 2, 1927. At that time the idea of template biosynthesis was too advanced for acceptance, and it was many years before it was used for experimental purposes. Today his idea of the chromosome as a single giant macromolecule is proved right, with the difference that it is not the giant protein strand but the deoxyribonucleic acid (DNA) that is found to be the basic reproducible template with genetic information. Although the structure of DNA was completely unknown in 1927, this does not reduce the theoretical importance of the basic principle of template biosynthesis. The idea of template biosynthesis as such proved a rational explanation of the possibility of auto-reproduction. These and many other works by Koltzov (17) provided the basis for physico-chemical cytology.

There was much other research during this period which provided the basis for new scientific approaches and breakthroughs. With more modern facilities I. P. Pavlov was able to carry out many new experiments on conditioned reflexes—the new field of physiology of the nervous system pioneered by Pavlov several years earlier. In 1923–1928, V. I. Vernadsky published his fundamental works about the biosphere; in 1924–1925, A. N. Bach, having established what was probably the world's first large Research Institute of Biochemistry, developed new ideas about the biochemistry of oxidation; and in 1927–1928, A. S. Serebrovsky did revolutionary work on the intragene variation of mutability. The first recognized theory of the origin of life on earth was developed in 1924 by A. I. Oparin. In spite of his later association with the pseudo-scientific ideas of Lysenko, Oparin's international reputation was not affected, for the simple reason that in the world of science he is known mainly as the author of the first scientific theory of the origin of life.

This list of achievements is far from being complete even for biology. However, the same kind of enthusiastic research was to be found in physics, mathematics, chemistry, geology, geophysics, and many other branches of science. New ideas were developed in technology and engineering. The country was ready for a new phase of technical and cultural development. The new experimental trends became evident in nearly all fields of intellectual life. The literature, art, music, and architecture of the 1920s has led to many new ideas, forms, and perfections. The young talents raised by the Revolution from popular grass roots together with those of an older generation of scientists and intellectuals proved the enormous fruitfulness of their cooperation. It was a brilliant start for Soviet science and technology, and one could have expected that with the end of postwar restoration and the beginning of prestigious new programs of rapid industrial development, Soviet science and technology would be a leading force in the reconstruction of industry and agriculture. This, unfortunately, did not happen. The situation for science and for scientists suddenly changed in 1929. The change was not only for science, of course: all groups and classes of Soviet society felt this dramatic turn in the history of the USSR, which its main architect, Stalin, once called "the Revolution from the Top," a turn for which Stalin took personal responsibility and which in historical retrospect must appear so terrible a tragedy.

Chapter 3

The First Wave of Stalin's Purges and Industrial Reconstruction (1929—1936)

In March, 1928, the apparatus of the State Political Administration (OGPU), which in fact was the secret police organization,[*] published materials about the "*shakhty* affair"[†] and accused large groups of engineers and experts in the coal mining industry of deliberate sabotage, of organizing accidents and explosions in the mines, maintaining secret criminal connections with former mineowners living abroad, and many other crimes. More than fifty experts were arrested. After a special show trial, eleven of them were condemned to be shot and others received prison sentences of varying lengths. This trial was used to start a campaign to increase the vigilance toward bourgeois technical and

[*] This organization changed its name in the following sequence: Cheka, (1918–1923), OGPU, NKVD, MVD, MGB, MVD, KGB (since 1954).

[†] *Shakhty* is the Russian word for "mines."

scientific experts, and to take new measures to increase the number of "red" experts in all branches of science and technology.

These new trends in science coincided with two major projects in industry and agriculture—the First Five-Year Plan of accelerated industrialization and the program of collectivization of individual farmers.

The "*shakhty* affair" was the beginning of several other similar "affairs." It is now known that they were falsified and that most of the "confessions" were obtained by the use of hard methods of "investigation" and torture. (See Roy Medvedev [18] for details.)

In April, 1929, in an address to the Party Central Committee, Stalin generalized the situation:

"Shakhtyites" are now ensconced in every branch of our industry. Many of them have been caught, but by no means all have been caught. Wrecking by the bourgeois intelligentsia is one of the most dangerous forms of opposition to developing socialism. Wrecking is all the more dangerous in that it is connected with international capital. Bourgeois wrecking is a sure sign that the capitalist elements have by no means laid down their arms, that they are massing their forces for new attacks on the Soviet regime [19].

After such directives from the highest authority, the vigilance of young "red" experts was directed against the older generation of scientists and engineers. Attempts to find "wreckers" in all branches of industry and agriculture proliferated, and soon "anti-Soviet," "bourgeois" elements had been "discovered" in almost all fields of science and technology. The faked "anti-Soviet" organizations which had been "exposed" included among their activists many famous scientists: the historian M. S. Grushevsky; S. A. Efremov, the vice-president of the Ukraine Academy of Sciences; the economist N. D. Kondratiev; A. V. Chaianov and A. G. Doiarenko, prominent experts in the agricultural sciences;

L. K. Ramzin, the director of the Heat Engineering Research Institute; many technical and scientific experts of the State Planning Commission; professors of universities and higher technical colleges; and members of many other research and state organizations.

All these forged and falsified proceedings and show trials had been designed to cover up Stalin's own mistakes and miscalculations. The collectivization of agriculture led to shortages of food commodities in cities and towns. An artificial situation of tension and emergency was convenient so as to force workers, farmers, and intellectuals to follow the hard line and to make possible the relocation of the working force in many new rural and Siberian construction sites. The measures against intellectuals were accompanied by more general repressions against farmers (kulaks) who opposed collectivization. Stalin also needed an emergency situation to solidify his extraordinary powers and to silence his critics.

In 1929–1930 large quantities of new machinery and technical equipment were imported from abroad. An accelerated program requiring the almost immediate use of this equipment in new plants and industrial sites which were still backward and sometimes uncompleted did really raise the rate of industrial accidents and failures. However, neither Stalin nor his associates wanted to take the responsibility for these accidents. The "bourgeois" experts became scapegoats, called upon to pay the heavy price of the leadership's failures.

The repressions against scientists and technical experts had direct influence on the very substance of Soviet science and technology. Condemnation, arrest, or execution of scientists—often prominent, often the leaders of certain schools of scientific thought or of theoretical approaches, or experts in a very specialized field of knowledge—usually encouraged their opponents and enemies to declare whole branches of science bourgeois, reactionary, capitalist, racialist, and so on. The members of these scientific schools were labeled by the name of the arrested leader (e.g., "Chazhanovists," "Kondratievists," etc.) or were branded with more general labels like "Mendelists," "Morganists," "Weismannists"—terms which came into use just at this time. The discussion in genetics, which is the best-known example of this kind of science "class struggle," was only one of many similar developments.

Many prominent Soviet biologists became targets for strong ideological criticism. S. S. Chetverikov, for example, who, as we mentioned earlier, was a founder of the Russian school of theoretical genetics, was arrested and exiled to a remote part of the Ural region. He was unable to continue his work there, and when he was rehabilitated, in 1955, about twenty-six years later, he was already too old to return to it.* The official press started a hostile campaign against N. K. Koltsov, A. Serebrovsky, and many other geneticists, accusing them of "idealistic" or even racial views on problems of heredity.

Party and state support for the younger generation of scientists was not only related to trust in its loyal "class" and ideological background. The older Russian scientific and intellectual elite was not large, but it was a highly educated group, with experience in theoretical "academic" research. The accumulation of knowledge needs time, and young scientists certainly felt the theoretical superiority of "bourgeois" professors. However, for a country which had just started to industrialize and had set industrialization as the main priority, theoretical knowledge was not considered very useful. The main emphasis was on the *practical utility* of scientific research, and young research workers, often of worker or peasant origin, easily advanced their claim to be a source of *practical* knowledge. They promised quick practical

---

* In 1959, Chetverikov received the special Darwin Prize, issued by the All-German Academy of Natural Sciences to commemorate the hundredth anniversary of the publication of Charles Darwin's classic work, *On the Origin of Species*. Among the recipients of the prize were such famous figures as H. J. Muller and J. B. S. Haldane. However, Chetverikov did not see the medal and diploma which accompanied the prize. He was already too ill and blind, and he died a few days after a telegram about the award was read to him.

results, revolutionary methods, and full support for the new industrial and agricultural projects, which had not always been really well balanced or properly programmed. This pragmatic approach to research became the dominating feature in the structuring of research networks, and the young and energetic representatives of the new generation began to occupy many commanding positions. They were not able to get such titles as "academician" or "professor," but these titles did not at this stage have the same respectability they had had in the past, or would have later. Many important administrative positions in experimental stations, provincial institutes, higher schools, and research departments of the People's Commissariats (Narkomnats) were held by young "practics," who often had a rather poor theoretical education. Had it not been so swift, extreme, and politically oriented, this shift to applied fields could probably have been useful.

In 1929, T. D. Lysenko got publicity for his "vernalization" methods (later proved inefficient). He was quickly promoted to the position of head of a large plant-breeding research institute in Odessa.

I have already discussed in a separate book (8) some aspects of the early stages of T. D. Lysenko's struggle with the established principles of genetics and the gloomy fate of many prominent biologists during this first major Stalin purge of "bourgeois" scientists. However, in 1929–1931 Lysenko and his supporters were not yet strong enough to win complete support for their "revolutionary" ideas in genetics and agriculture. The arrest and imprisonment of some prominent biologists and experts in agricultural sciences (S. S. Chetverikov, S. K. Chazhanov, N. D. Kondratiev, and some others) were not directly related to the beginning of the "Lysenko affair." The disputes, although very sharp in language, were mostly around theoretical problems of inheritance of acquired characteristics.

The organized repression of intellectuals in technical fields quickly spread to all academic and scholarly communities. During the summer of 1929 the Academy of Sciences of the USSR had already become one of the main targets of the purges. At that time the Academy had not yet moved to Moscow, and the main institutes, as well as the presidium, library, publishing house, and many other administrative units were situated in Leningrad. The Leningrad Party Committee established a special government commission to investigate the Academy, and this commission found it to be the "center for counterrevolutionary work against Soviet power." The official history of the Academy of Sciences of the USSR (1) does not say anything about the purges of 1929, which seriously interfered with work of the Academy and which really ended the Academy's comparative independence. Three academicians were arrested (S. F. Platonov, N. Lichachev, and M. K. Liubarsky), and hundreds of other members of the staffs of the presidium, library, and publishing house, as well as senior research workers of the institutes, were either arrested or dismissed (20).

The election of new academicians in 1928 had probably been the starting point of political controversy over the Academy system. Before 1928 the Academy was apolitical, and old academicians resisted electing scientists who were also members of the Communist Party. In 1928 the number of vacancies had almost doubled, and the elections were publicized in the central press. Great pressure had been brought upon the Academy to elect eight members of the Party (among about forty new academicians) as academicians. This group soon started to lead the purges in the different branches of the Academy. The fact that more than two hundred senior research workers in different institutes and administrative bodies had been arrested, and many more dismissed, is described in the recent official history of the Academy (1) as the "strengthening of Party influence on all aspects of life of the Academy of Sciences. The leadership of many institutes, museums, commissions . . . had already in 1930 been strengthened by Marxist cadres" (pp. 288–289).

At the same time industrialization and collectivization of agriculture made a sharp increase of highly qualified professionals and experts in all branches of Soviet economy urgent. The need for greater efficiency in science and technology was partly related to the increased import of foreign machinery and equipment for science, industry, and agriculture. The new technology was not used properly, and was often simply spoiled or damaged by the inexperienced young "red" technical generation (the loss of imported foreign equipment was about 20 percent during the First Five-Year Plan, which was completed at the very end of 1931, before the scheduled date). This situation led, after 1932, to a temporary reduction of the ideological struggle within the scientific and technical communities. Some experts were released from prison; others, like Professor Ramzin, were allowed to continue their research work in prison. (Ramzin, who developed while in prison a new, revolutionary model of boiler construction for heat engines, was later released and even given an award.)

Rapid industrial growth and the import of modern foreign technology during the First and Second Five-Year Plans exposed the backwardness of many branches of Soviet technology and science. In an attempt to overcome this, during this period many experts from abroad were invited for technical and research assistance. The great economic world crisis and depression of 1929–1933 helped the Soviet government to employ many thousands of foreign experts from Germany, France, the United States, and other countries. Special small villages of "foreign experts" were constructed in the main centers of industrial development, and in these "villages" foreign "bourgeois" experts enjoyed a much higher standard of living and many more privileges in consumer goods and services than their Soviet colleagues doing the same work. Many foreigners were invited to work in the new research institutes. At the same time there was a drastic reduction in travel abroad by Soviet scientists.

The Academy of Sciences of the USSR and the Institute of the History of the USSR have recently published a detailed work about the international contacts and cooperation of Soviet science, technology, and culture (21). However, the book restricts itself (with many omissions) to the period between 1917 and 1932. One might think that this was only the first volume of a larger work and that the periods after 1932 would later be considered as well. The truth is much more tragic—after 1932 almost all contacts of Soviet science with foreign countries were broken. Many academicians and other prominent figures who, by the nature of their work, depended on foreign trips—for example, N. I. Vavilov, the geographical botanist and geneticist, or A. N. Severzov, an evolutionary and geographical zoologist—lost their previous freedom to travel on expeditions. The same happened to writers, poets, actors, and other intellectuals.

Some personal tragedies (for example, the suicide of the famous Soviet poet Vladimir Mayakovsky in 1930) were partly connected with the sudden end of the freedom to travel abroad. A few desperate scientists tried to escape by crossing the border illegally. Professor G. Gamov's story is a good example of the fate of such attempts.

Gamov, a well-known physicist, who was afraid of possible arrest, twice tried to escape from the USSR by the way which was still comparatively easy before 1929—illegally crossing the border.

Before 1929 border control was rather primitive and the whole border was not always guarded. However, after 1928–1929, when not only forced collectivization but also the liquidation of the New Economic Policy had begun, thousands of people (richer farmers, former owners of small shops and other private enterprises) tried to escape this way. Border control became much more intensive. For his first attempt to escape, Gamov selected the distant mountain border with Afghanistan, but he was caught by soldiers. He explained that he was a mountain climber and sportsman, and this excuse was accepted. The next time, when he was at a Crimean Black Sea resort sanatorium, he tried to cross the Black Sea into Turkey in a small boat. However, the Coast Guard gunboats already patrolled the territorial waters. Again, the explanation that his hobby was sailing was enough, and, surprisingly, neither attempt aroused suspicion. And that is why Gamov was later permitted to attend an international congress abroad, from which he never returned. He settled in the United States, and was famous among biologists at the beginning of the 1950s for his theory of transfer of information between nucleic acid and proteins by a special code formed by nucleotide sequences—a theory which proved very successful, and which became a working theory for many experimental approaches to the establishment of the genetic code.

However, all these repressions did not stop the general growth of Soviet science.

The Academy of Sciences of the USSR and other academies almost tripled their membership, and the number of students at universities and at technical medical and agricultural higher schools increased between 1929 and 1937 by five or six times. In 1913 there were 298 universities and other higher schools and research institutes in Russia; by 1937, in the USSR, this figure had increased to over 2,000 (it was 2,359 just before World War II), and the total number of research scientists had grown to about 150,000.

However, the increase in the number of research scientists was not a sign of perfectly healthy growth. The ideological struggle, which had begun so violently in 1929–1931, did not stop. It was reduced after 1932 and was transformed into virulent discussions, accusations, and counteraccusations. "Bourgeois" scientists and their followers were again tolerated; their knowledge was necessary for the many new industrial and scientific projects and for the proper assimilation of imported equipment. But the notion that bourgeois science is *limited* was still the official dogma. One of the assumptions of early Marxism was that capitalist economy restricts scientific and technical progress to a certain level, and this was still considered to be true. This level had to be reached by the new *Soviet* science and technology, but after this crucial point the socialist "proletarian" science and technology would lead the world technological and scientific revolution. This was the official belief of the Party leaders. It was also the belief of many young scientists and engineers, who thought that the methodology of dialectical materialism would help them automatically, its pattern of reason helping to reveal new laws of nature which were closed to bourgeois scientists working empirically.

This crucial point had certainly not yet been reached in 1937 on the eve of the decisive Third Five-Year Plan, and to fulfill the targets of this plan the assistance of foreign knowledge and "bourgeois" science would still be necessary. However, the political events of 1937, which are known in history as the "Great Terror," made an irreversible rampage on Soviet science as well. The explosion of terror was destructive for the country. It was even more destructive for Soviet science.

Chapter 4

Stalin's "Great Terror" (1936–1940) and "The Great Patriotic War" (1941–1945)

The Great Terror began in 1936 and reached its culmination in 1937–1938. Several million people were arrested and imprisoned and more than half a million were executed in what constituted a unique phenomenon in world and Russian history. I do not plan to discuss here the full impact of this terror on the Soviet scientific community.[*]

[*] See Roy Medvedev (18) and works by other authors (22, 23). Detailed descriptions of the effect of the terror on biology, genetics, and agricultural sciences are given in my book (8) and in David Joravsky's (24). Joravsky offers a list, albeit incomplete, of prominent Soviet scientists (mostly biologists) who were arrested during this period. This list includes about one hundred names. Roy Medvedev also gives the names of more than a hundred famous scientists in all scholarly fields arrested during 1937–1938.

Nonetheless, this constitutes only the tip of the iceberg. Scientists and technical experts with international reputations like the academician Lev Landau, later awarded the Nobel Prize for Physics; academician Nikolai Vavilov, former president of the Academy of Agricultural Sciences; Professor Georgii Kar-

The full list of arrested scientists and technical experts certainly runs into many thousands, because the arrests usually set off a chain reaction. The leaders of scientific schools or the heads of design project bureaus were often arrested, together with their whole research teams. In the natural sciences, the mass arrests of prominent scholars led, as a rule, to the emergence of substitute figures—charlatans of the T. D. Lysenko type, who were declared leaders of the new "proletarian" versions of one or another scientific trend. The arrests and executions did not, however, reduce the total *number* of members of the scientific elite. After the election of new academicians in 1939, for example, the Academy was larger than at any time in its history. In the list of new academicians there were not only frauds like Lysenko, but some good scientists as well (like mathematicians S. Sobolev and A. N. Kolmogorov, physicists P. L. Kapitsa and V. A. Fock, and many others). In all there were 56 new full members and 102 new corresponding members. Since the natural sciences do not have an immediate application to a country's economy and agriculture, the destructive effect of the repressions of 1936–1938 were felt only later on, but then for many years to come. On the other hand, the mass arrest of experts in technology had an immediate impact on the country's development and on the quality of industrial and military projects. Industrial technology already

petchenko, the famous geneticist; A. N. Tupolev, the best designer of Soviet aircraft—these were only a few of the most prominent.

It is now generally assumed that the destruction of biology, genetics, and agricultural sciences in the Soviet Union which reached its peak during the years of Stalin's terror was the most bizarre chapter in the history of modern science. The damage in technical sciences became evident in a few years, and the regime tried some remedies in the form of special "prison-research centers," which I describe later. But in biology and agricultural and medical sciences, the substitution for genuine research and development by some pseudoscientific network was not so visible, and the abnormal situation in some cases existed for more than twenty years. The distortion of theoretical and applied sciences initiated in the mid-thirties was so serious that its consequences are responsible for the backward condition of many practical fields of research today.

in use was not much affected; electric power stations continued to work, auto and tractor plants went on producing cars, trucks, and tractors, the aircraft industry continued to build 1936-model planes. But the development of *new technology,* and indeed scientific and technological progress in general, was greatly delayed and in some areas simply halted. The international connections of Soviet science and technology were completely severed. Foreign experts, voluntarily or forcibly, left the country, while in some cases they were arrested, together with their Soviet colleagues.

These years of terror created a kind of technological vacuum in many vitally important branches of industry, especially in those areas where continuous technological innovation is an absolute necessity. The best example of this is the aircraft industry. The Soviet Union in 1927–1936 had created a modern aviation industry. Many very good technical centers were established where new models of military and civilian aircraft were designed. In 1935–1937, Soviet pilots broke many world records in aviation. Paradoxically, the elite of the aircraft industry suffered more from Stalinist repressions than did that of any other branch of technology. The finest designers of planes—A. N. Tupolev, V. M. Petliakov, V. M. Miasishchev, V. A. Chizhevsky, and many others—were arrested and imprisoned. Development of new models of aircraft was virtually halted. The many new engineers and experts who were appointed to direct technological development and to create new models simply did not have enough experience, knowledge, or talent to fill the gap. Their "proletarian" background was not an adequate substitute for the knowledge and technical imagination which the older generation possessed.

In botany, zoology, or genetics the country did not feel any immediate effect as a result of the elimination of scientific leaders and their replacement by mediocrities, charlatans, or simply young, inexperienced scientists. But in the sensitive areas of complex technology the results were immediate. Unfortunately, the damage to Soviet technology, especially to military technology, occurred just a few years before the beginning of the inevitable war with Hitler's Germany.

When the war started, in June, 1941, the USSR had numerical superiority in tanks and aviation, but Soviet planes, rifles, and most tanks were not at a 1941 level. The qualitative superiority of German military machinery was evident; it was this technological superiority of the German military industry that was mainly responsible for the rapid defeat of the Red Army during the summer of 1941. When Soviet soldiers saw how a few German Messerschmitts could bring down dozens of slow-moving Soviet planes, they could only weep in despair. In two years, however, the situation changed. By 1943–1944 the Soviet military technology had gained new strength.

The progress made was peculiar to the Russian situation. Technical and military assistance from Soviet allies, Britain and the United States, was a contributing factor. Even more important was the fact that German science and technology suffered from something like the same kind of "ideological" and "racial" damage before the war. In the case of Germany, however, many hundreds of the country's best scientists who were harassed and dismissed or temporarily arrested emigrated abroad. By 1940 they were working mainly in the United States and in Britain. The Russian technical experts remained available, albeit behind bars, except, of course, for those who had been shot. Their knowledge and experience could be used. They could be released or put to work in special prison research and technical centers created for the purpose. Prison design and engineering bureaus, prison research institutes, prison scientific complexes, came into existence not long before the war. The work done in these places, plus the work of free experts, soon changed the qualitative balance in favor of the Soviet military machine.

The rules in these prison research centers were very strange. The chief engineers, though prisoners, often headed large teams in which some experts were free, not prisoners themselves. After the long working day such free employees could return home to their families; their superiors were taken to their prison cells. During the war many prisoners were promised release for completing successful projects, and this constituted a strong incentive for hard work. Violations of rules and project failures could lead to the transfer of prisoners from these "privileged" places into ordinary prisons and labor camps.

The author of the *samizdat* work "Tupolevskaya Sharaga," Professor G. Oserov (25), describes his arrival at the main Moscow "aircraft prison center" as follows:

> We were taken to the dining room . . . heads turned to our direction, sudden exclamations, people ran to us. There were so many well-known, friendly faces. At the tables we can see A. N. Tupolev, V. M. Petliakov, V. M. Miasishchev, I. G. Neman, S. P. Korolev, A. I. Putilov, V. A. Chizhevsky, A. M. Makarov, and many others—the elite, the cream of Russian national aircraft technology. It was impossible to conceive that they had all been arrested, that they were all prisoners— this meant a catastrophe for Soviet aviation!

A. N. Tupolev later told G. Oserov that these people were only a fraction of all the Soviet experts who had been arrested. When the research center was established in 1938, a search began in labor camps and prisons throughout the country for experts to man the center. Too many, however, were already dead or under sentence of death. Korolev was, for example, the only person from the special rocketry research unit who had survived.

In Tupolev's *sharaga,* or prison, there were three special departments engaged in the development of three types of military aircraft: high-altitude fighters, long-range bombers, and dive-bombers—categories of planes which the German army already possessed but which were not in the Soviet air force arsenal. A new weapon that played a very important role in many wartime battles, and especially in the defense of Moscow in December, 1941, was the famous multishell gun attached to tracks that fired simultaneously about a dozen reactive jet shells, sometimes with the explosive prototypes of napalm. In the Red Army these guns were called "Katiushas." * The main designer of the guns, which represented a considerable breakthrough in military technology, was Georgy Langemak, who had been arrested and sentenced to death in 1937, together with his whole engineering group. One of his assistants, A. G. Kostikov, who had somehow remained alive, later resumed the work, organized industrial production, and made the gun available for the Red Army at the very beginning of the war.

Among the leading physicists arrested, probably the most dramatic case was that, in 1938, of L. D. Landau, one of the greatest luminaries in Soviet physics and a future Nobel Prize laureate. Landau was senior research worker of the Institute of Problems of Physics, specially created for P. L. Kapitsa in 1934. Kapitsa tried to obtain the release of his friend, although the procurator-general had already informed him that Landau had been sentenced as a "German spy." When this attempt to free Landau failed, and after a short meeting with Landau in prison, Kapitsa took a desperate step. He presented Molotov and Stalin with an ultimatum: if Landau was not released immediately, he, Kapitsa, would resign from all his positions and leave the institute. Kapitsa had, at this time, a great international reputation and the government considered him to be very important for some key research projects, partly related to atomic physics. It was clear that Kapitsa meant business. After a short time Landau was cleared of all charges and released.

Although the repressions undoubtedly delayed the development of Soviet science and technology and created a gap between Soviet and Western research, they did not halt scientific progress entirely. The enthusiastic start made in the 1920s had a certain impact on the development of science in the 1930s as

---

* The modern version of this system is known in the West as "Stalin's organ."

well. Though the number of scientific and technological researchers swelled, there was no parallel growth of productivity. Not many scientific discoveries or technological breakthroughs were made in the 1930s. The picture in other areas of intellectual life (literature, theater, philosophy, etc.) was even bleaker. Science and technology suffered mainly from the loss of highly creative individuals, while in literature, for example, there was additional severe psychological damage; fear completely halted the enthusiastic experimentation of 1922–1928.

The Academy of Sciences of the USSR was transferred in 1934 from Leningrad to Moscow and was made directly subordinate to the government (decree of April 25, 1934). To increase the political orientation of the Academy, which at this time became the most important scientific institution in the nation, a decision was made by the Party Central Committee and by the government on February 7, 1936, to merge the Academy of Sciences and the Communist Academy in Moscow. Until now several institutes of the Communist Academy had been included in the Academy of Sciences network (the Institutes of Economy, History, Philosophy, Soviet Law and State, World Politics, and some others).

In examining scientific development during this period one must mention some very important work and discoveries that received world recognition. Scientific progress in the Soviet Union was largely in physics and mathematics, which had suffered comparatively less from repression than, for example, biology or technology.

In 1937 the physicists I. E. Tamm and I. M. Frank found an explanation for the so-called "P. S. Cherenkov's effect," the fluorescence of some liquids irradiated by gamma rays. This discovery, which later proved important for laser technology, was awarded the Nobel Prize in 1958. (Cherenkov, Tamm, and Frank shared the prize.) In 1937 the Second All-Union Congress of Physicists showed that there had been a significant development in the field of atomic physics in the USSR. At this congress, Professor Igor Kurchatov gave several important papers about induced radioactivity. Kurchatov was the head of the team which had designed several types of accelerators of elementary particles and built the first cyclotron in Europe (1937). K. A. Petrzhak and G. M. Flerov, Kurchatov's research assistants, made a fundamental discovery in 1939: the fission of uranium nuclei. The same discovery had been made independently a few months earlier by O. Hahn and F. Strassmann in Germany.* The first system for the industrial preparation of liquid helium was made in Kapitsa's research institute.

In 1937 the USSR organized the first successful long-term research expedition to the North Pole, and a team of pilots made a revolutionary flight from Leningrad to the United States via the North Pole route. The flight was made by the plane ANT-25, designed by A. Tupolev, who was arrested the same year.

It was natural that during the war the entire scientific community, including most theoretical research institutes of the Academy of Sciences of the USSR, worked mainly in military-oriented fields. The institutes were under an obligation to assist military technology and their contribution was very substantial. For example, as soon as the German air force started its air raids on Moscow and other major cities, Kapitsa's Institute of Problems of Physics received an order to develop a safe method for defusing unexploded bombs. The method was developed in the space of one week and proved efficient. Kapitsa also designed the industrial system for the production of liquid oxygen, hydrogen, and other gases.

Research teams of physicists and mathematicians improved Soviet multichannel guns with jet shells ("Katiushas") and reduced the randomness of shell distribution. Resources were explored in the Urals, Siberia, and other eastern parts of the country—work which resulted in the discovery of several new oil fields.

---

* The war, which had started in Europe in 1939, made communication with German physicists extremely difficult.

One can mention several dozen technical and applied scientific projects that were developed during the war. It was not uncommon for "free" research workers, sometimes academicians, to cooperate with their imprisoned colleagues in prison institutes. The falsity of the accusations against scientists and technical experts was, in many cases, absolutely clear, but they were nevertheless kept in prisons. Some tragic paradoxes obtained during the first two years of the war when the advances of the German army often made the evacuation of important research and industrial centers necessary. But the evacuation of prisoners was seen as being too difficult and complex a problem.

To simplify matters, many prisoners were simply shot. Others died from malnutrition and exposure; academician Nikolai Vavilov died in the Saratov prison in 1943, G. D. Karpetchenko died in a prison camp in 1943, and there are many other examples. The execution of political prisoners (former "Mensheviks" and members of other parties), while other younger and more "productive" people were evacuated from front-line cities, has been described by S. Gasarian (26). The security forces were more prudent, however, when it came to evacuating technical experts—for example, prisoners who were working in special groups to develop new types of fighter aircraft. One such group was evacuated, together with the aircraft plant, to the Siberian town of Omsk. When the group arrived there in the early fall of 1941, it happened that the city had no prison available to keep "enemies of the people" and "spies" under guard. Had they not been aircraft engineers, they could probably have expected execution. But the aircraft plant was too important. An exchange of telegrams with Moscow resulted in an extraordinary event. The large group of aircraft experts that had been arrested in 1937 and imprisoned for long terms as "spies" were assembled in the office of the NKVD general, A. N. Kutepov. The general took a piece of paper and started to read: "The Soviet Government, taking into consideration the successful work of the experts listed below on the development of new aircraft models, has decided to free them and to release them from under arrest" (25).

They became free people again and the new aircraft plant soon began working. The machines were assembled and put into operation under the open sky—production was more important than building the walls. The plant was in full operation and the new fighters very nearly ready to be sent to the front when construction of the walls began. From 1941 to 1945 the average speed of Russian fighters was increased about 100 km. per hour above the prewar level (1). The speed of German fighters remained unchanged throughout the war.

The war was a great tragedy for the USSR; twenty million were killed—most of them young, healthy people. Still, the war produced some measure of improvement in conditions of life. In order to unite all citizens in the struggle against Hitler's army, even organized religion was encouraged. Many of the churches that had been closed in 1921, 1929, or 1937 resumed services.

When the highly dogmatic former personal secretary to Lenin, E. Stasova, tried to argue with Stalin about all this "new mysticism," Stalin replied coldly: "We cannot raise the morale of the people and win the war by Marxism-Leninism alone" (27). When the war was over, however, pragmatic tolerance toward religion by the "Great Leader" was also over. The most bizarre chapters in the history of Soviet science were to follow in the postwar and cold war period of 1946–1953.

Chapter 5

The Postwar Period (1946—1953)

The creation of a parallel network of special prison research centers before the war was an attempt to compensate for the damage caused by the arbitrary arrests of thousands of technical experts and scientists. The Great Terror slowed down the development of military technology to a dangerously low level, and Stalin had no choice but to put even sentenced experts and scientists to work as professionals. Engineers and scientists were collected from many different camps. Some of the prisoners had been picked up when they were close to death from hard work and malnutrition. Now the famous S. P. Korolev, designer of intercontinental missiles and the first earth satellites (*Sputniks*), was picked up from one of the most terrible Arctic Kolyma camps, where he worked as a miner. (My father, who was arrested in 1938, died in the Kolyma mines. He was a professor but, unfortunately, neither of physics nor of chemistry, but of philosophy.) Of course, hundreds of experts either were not found or had already died when this kind of research institute came into existence. (The novel *The First Circle,* by A. Solzhenitsyn, gives a picture of one such research prison institute.)

This prison research network proved itself very efficient during the war, and many military designs which changed the qualitative balance in favor of the Soviet army after 1943 (new tanks of T series, new planes PE-2, TU-5, 1-5, some new artillery systems, and even new locomotives for railroads) had been developed by imprisoned scientists and technical experts. Some of them (including A. Tupolev and S. Korolev) were released after the war as a reward for their successful work.

The postwar period was associated with new waves of repressions against intellectuals. In addition, many thousands of German, Hungarian, and Romanian scientists had been either taken prisoner during the war or deported from occupied territories of their countries to work in the research prison centers.

The latter years of the war proved the superiority of the Soviet army over the German army, but they also proved the inferiority of Soviet technology compared to American technology in military equipment. Not only were the American Studebakers which were sent to Russia as part of a lend-lease program much better than the Soviet-made trucks of this type, but American naval and aviation equipment, and anti-aircraft defense systems, and much else in the technological-military area were better than the Russian designs. Not long before the end of the war one U.S. bomber, a Flying Fortress type, was damaged (or ran out of fuel) during a raid on Japan and made an emergency landing in the Russian Far East. It was immediately brought to experts who had been given Stalin's personal order to study all details down to the last screw. The report prepared for Stalin indicated that the American bomber was better from all points of view than the Russian strategic bombers. (This story was told to me in 1958 by Golovanov, former Marshal of Aviation, then in retirement.)

The first atomic-bomb explosions made the military and technical superiority of the United States even more evident. This

was a definite embarrassment for Stalin. One of the main principles of Communist ideology was Lenin's erroneous assumption
that capitalism is limited in its capacity to develop technology and
science and in its last stages will inhibit scientific progress and
the productivity of labor. That only socialism would be able to develop the *full potential* of technology and science was not proved
by the reality of the American atomic bomb. This situation forced
Stalin to acknowledge the gap between U.S. and Soviet science
and technology and to make in 1945 his famous directive: "If we
[the Party] help our scientists they will be able to reach and to
surpass the achievements of science abroad."

The cold war situation which started in 1946 made cooperation
between the Soviet Union and Western countries impossible.
The directive within the Soviet Union was clear—build a powerful scientific establishment able to pursue *all* problems and develop *all* technology which the opposite part of the world is capable of doing, and make Soviet science not just equal but superior.

From 1946 practically all branches of military-oriented science
and technology received the highest state priorities (which improved the position of general science as well). The Academy of
Sciences got new powers and was able to organize several dozen
new research institutes. Financial resources for science were
increased sharply. The average salary for scientists was doubled
or tripled, and in a country where food and consumer goods were
still rationed, scientists found themselves in the highest privileged group.

Almost half of the western part of the Soviet Union was in
ruins, and the farmers of many destroyed villages lived in dug-
outs (*zemlianki*) on the sites of their war-burnt homes, but scientists suddenly became the privileged elite of the country, their
living standards having been raised much higher than the prewar level. The new institutes multiplied like cells in a culture,
and almost all demobilized soldiers who had a secondary education (equivalent to the American high school) were absorbed by
the enlarged network of higher technical schools and universi-

ties. The number of students, which was 817,000 just before the war, reached more than 1,500,000 in 1948–1949.

After the war the Soviet Union became the leader of the Communist bloc—in science as well as in politics. The underlying motive was clear—to create a parallel but more productive and more successful *socialist* science which would represent all branches of science abroad and be the background for the socialist economy. The different areas of research began to take the shape of a pyramidal hierarchy—the authoritarian type of state and party rule carried over to science. Each specific area of research was headed by a "leader," who assumed not only the administrative powers but the theoretical and scientific leadership as well. Each area of research had to prove its "socialist" specificity, a distinction which would separate it from bourgeois "idealist" science. Each field of the natural sciences had to be based on the principles of "dialectical materialism" and to use as its fundamental background the "great ideas" of Marx, Lenin, and Stalin.

In the most important areas of research, which had special priority, the scientific leader received full support from the government and the Party and usually had a so-called "open" account in the state bank (in rubles as well as in foreign currency). This meant that such research leaders could spend as much as they wanted to establish new laboratories, institutes, and special programs.

Although it is not easy to trace the results of this policy in all fields, it was certainly a disaster in biology and agriculture, where T. D. Lysenko was the leader in charge of all activities—theoretical and practical.

The Lysenko affair represented in many ways Stalin's own interference in the substance of scientific research. Stalin was convinced that he understood the main dialectical principles of biology and evolution. He also considered himself an authority in some other scientific fields, not to speak of political science, linguistics, and philosophy. However, he did not try to interfere in the research of experimental physics, and the large group of physicists collected by the academician Igor Kurchatov enjoyed not only unprecedented power to mobilize all available resources for atomic projects, but even the freedom to avoid "political education." Stalin once told somebody who was responsible for political and ideological education in science, "Do not bother our physicists with political seminars. Let them use all their time for their professional work."* As a result, Soviet physics was much more successful than biology.

When the U.S. air force dropped two atomic bombs in September of 1945, Kurchatov's research group already knew the theory of reactors and the possible methods of nuclear explosion. However, the available supplies of uranium were too small to start real work and no other facilities were ready. The site for a big research center for atomic energy in a northern suburb of Moscow was selected at the very end of 1945. Only one year later, on December 25, 1946, the first large experimental reactor was tested there. This reactor was able to accumulate plutonium, but it was not big enough to accumulate the amount necessary for military use.

During the next year, 1947, the construction of a big military reactor was completed and the assembling of several other reactors already started in distant parts of the country, usually far from Moscow. In most cases, these were secret locations where special prison research centers were also situated. Kurchatov was the research head of the whole project; Lavrenty Beria was the administrative officer, with unlimited resources of slave labor of all qualifications. All German physicists in the occupied part of Germany were transferred to these camps.† Some important advances were made by the German prisoners. One of them was

---

* This "instruction" of Stalin's was often mentioned by political lecturers in 1973 to explain academician Andrei Sakharov's anti-Marxist and anti-socialist attitudes.

† In two main biographies of Kurchatov (and the Soviet atomic project as well), this side of the story was never mentioned (28, 29).

later awarded the title "Hero of Socialist Labor" and was released to head nuclear research in the German Democratic Republic. However, he soon escaped into West Germany and most likely is the only Western scientist holding such a high title of honor from the Soviet Union.*

From Kurchatov's official biographies we can conclude that the building of several large plutonium-producing reactors started simultaneously in 1947. Kurchatov had personally headed and inspected all building projects and works, and was in charge of the complex processing plant that separated plutonium from uranium blocks and produced large quantities of different kinds of nuclear radioactive waste. The Soviet government's official statement about possessing the "secrets" of the atomic bomb had been made at the end of 1947, which meant that a method of separating plutonium from uranium was known. It was developed in Moscow, where a small experimental reactor was operational. Reprocessing of plutonium from the first big military reactor had begun at the end of 1948 or the beginning of 1949. The design of the actual bomb was a totally different project, and Kurchatov probably was not directly responsible for this design. It was in the hands of the military branch of the field, and neither the names nor the history of this part of the Soviet nuclear project were disclosed or declassified for writers.

The first nuclear explosion was made in the USSR on September 23, 1949. The deliverable bombs were tested in 1950. (I think that 1949 was the deadline set for Kurchatov's team. Stalin celebrated his seventieth birthday in December, 1949—which was programmed as a great event for the whole Communist bloc.

* I have omitted this scientist's name because all titles and prizes for nuclear research were given secretly, and it is possible he has not informed his Western colleagues about his Gold Star and the Order of Lenin which accompanied the title. This story was told to me by the famous geneticist and radiobiologist N. V. Timofeev-Resovsky, who had lived in Germany since 1926 and had been captured there in 1945. From 1946 until 1955 he worked in one of the biggest labor-camp research centers closely related to nuclear projects in the South Urals. Some details about his work there are discussed in a special chapter of my earlier book (9).

If the atomic bomb were not available, this celebration would not have been so glorious.)

The other scientific fields in the USSR during the postwar period were not so successful. Pseudo-scientists like Lysenko, O. Lepeshinskaya, M. Olshansky, and many others still dominated official "Soviet science" by occupying the most prominent positions after the war. However, the nuclear industry had a great influence on the general development of Soviet science. Radioactive isotopes and radioactive synthetic materials became available for experimental work in chemistry, biochemistry, physiology, and many other branches of science. These new methods transformed research possibilities more than anything else in the history of science. The use of isotopes was an important factor in the subsequent development of a number of sciences, and resulted in significant growth in many laboratories. The work with isotopes and radiation demanded new knowledge, in physics, biophysics, statistics, and so on, and thus required a departure from the primitive conditions and experimental anarchy that had prevailed in many laboratories in pre-nuclear times.

When radioactive isotopes first became available for research work in biology, very few people were aware of their dangers, and no strict safety rules existed.

I remember well how I began my own work with radioactive isotopes. It was in 1951 when I was a young junior research worker in the Department of Agrochemistry and Biochemistry at the Timiriazev Agricultural Academy in Moscow. My professor at that time, A. G. Shestakov, invited me and a friend to his office. He said, "Listen, boys, I have twenty millicuries of radioactive phosphorus. We must do something with it." He put his hand into the inner pocket of his jacket and extracted the glass ampoule with some liquid inside it. I took the ampoule with my bare hands and put it into my pocket. Neither our professor nor we had had any previous experience with isotopes. (Now, of course, I know well that the thin glass ampoule was no protection from P^{32} radiation and that 20 mc. was a dangerous dose for local irradiation.) At the department we did not have the equipment to count the radiation level, but we nevertheless started our first experimental work with a simple project. We decided to compare the distribution of the isotope in different parts of some plants, using autoradiography on large X-ray film sheets to measure the comparative activity. Although it was a simple project, it was much more exciting than anything we had ever done before. In a few years I was more or less considered an expert on working with different isotopes for biochemical research and began teaching in this field. In 1957, I had the privilege of presenting my paper at the plenary session of the First International Congress on Radioisotopes in Scientific Research, which took place in Paris (30). I was a member of a large Soviet delegation. This was my first and only trip abroad before 1973, when I was permitted to visit the National Institute for Medical Research in London, but not permitted to return home. (My Soviet citizenship was "canceled" in 1973 by special decree of the Supreme Soviet Presidium. There are some invisible connections between the dangerously radioactive ampoule which my late professor took out of his pocket and the actions of the Soviet Embassy officials in London who confiscated my Soviet passport in August of 1973.)

A victory in the Second World War did not end the military orientation of Soviet science and technology. The system of prison research centers continued, although many prisoners who made important contributions to military technology were released. A. N. Tupolev, S. Korolev, and V. M. Petliakov were among those released. The cold war and the development of tense relations with the United States and Western Europe concerning the postwar fate of Eastern Europe made the deterioration of East-West relations inevitable. This meant that it was necessary for the Soviet Union to try to match former allies in three decisive fields of military science and technology: atomic bombs, supersonic aircraft, and rocketry. The aircraft industry was already well developed and powerful, but the nuclear and rocketry programs were practically new—the industrial base for them had not been available in 1945.

Many of the nuclear and space science centers—now attractive scientific towns near big industrial cities (the most famous around Moscow are Dubna, Obninsk, Chernogolovka, and Pushchino)— were founded in 1946 as research prison camps, large complexes isolated by barbed wire and blessed into existence by Beria, where Russian prisoners worked there side by side with deported German experts, engineers, and scientists. In East Germany, the Soviet army captured a great quantity of German military equipment, machinery, plants, patents, plans for German nuclear projects, and dozens of different types of military rockets.

A special rocketry center was established in 1946 in a deserted part of Bashkiria (not far from the Caspian Sea) to study and to test these German rockets. S. P. Korolev, now a free man, was appointed to head this program. Russia's first military rocket was designed by Korolev in 1947. At that time the mathematical theory section of rocketry engineering was headed by M. V. Keldysh, who later became president of the Academy of Sciences of the USSR.

Kurchatov's biographers (28, 29) always referred with admiration to the speed which was typical of all construction for the nuclear industry. The first reactor to produce plutonium was started in Kurchatov's laboratory in Moscow at the end of 1946, but large industrial reactors of this type were already under construction.

> Far from Moscow in a picturesque location the new town was already taking shape. Chemistry buildings, supplementary constructions and large blocks of industrial reactors were close to completion. In January, 1947, Kurchatov sent there his close assistants to consult final works. During the fall of 1947, Kurchatov himself arrived here to start the reactor. The large town was born with thousands of workers, technical experts, engineers [28, p. 70].

There is no doubt that these were *special* labor camps, camps where "free" scientists worked together with their imprisoned colleagues.* The industrial complexes were within Beria's Gulag empire, and they could be built cheaply and quickly because normal housing, which a "free" labor force would consider a necessary preliminary condition for working, was not included. "Free" workers also have families; they need not only living space but schools, kindergartens, hospitals, and a lot more. Prisoners need only wooden barracks where hundreds of them can be accommodated in small spaces with double-tiered bunks. Later, when the main industrial objects were ready and had been put into operation, the prisoners began the construction of real towns for "free" workers—physicists and their families who arrived at the new place for *permanent* work.

The camp could be destroyed later and prisoners sent to a new construction site. This was the typical pattern of the swift development of the new centers for the nuclear, rocketry, and aircraft industries.

Andrei Sakharov, who since 1950 had worked on nuclear projects under this system, described in his book *Sakharov Speaks* (31) how he started to realize its immoral implications.

In 1950 our research group became part of a special institute. For the next eighteen years I found myself caught up in the routine of a special world of military designers and inventors, special institutes, committees and learned councils, pilot plants and proving grounds. Every day I saw the huge material, intellectual, and nervous resources of thousands of people being poured into the creation of a means of total destruction, something potentially capable of annihilating all human civilization. I noticed that the control levers were in the hands of people who, though talented in their own way, were cynical. Until the summer of 1953 the chief of the atomic project was Beria, who ruled over millions of slave-

* According to recent disclosures, 70,000 inmates of twelve prison camps worked from 1945 to 1948 to build this "picturesque new town." This area was also the site of the nuclear disaster of the winter of 1957–1958, described in Chapter 6 and Appendix II.

prisoners. Almost all the construction was done with their labor. Beginning in the late fifties, one got an increasingly clearer picture of the collective might of the military-industrial complex and of its vigorous, unprincipled leaders, blind to everything except their "job" [p. 31].

Sakharov's special sensitivity to the dangers of nuclear tests certainly was related to his own brilliant design, which made the Soviet hydrogen bomb project so successful. At first the Soviet program was close to the American one. The Americans exploded their device on November 1, 1952, on one of the Marshall Islands. It was, however, about a sixty-five-ton complex, which was not deliverable by rocket or plane.

The Soviet H-bomb was exploded in August of 1953 on Wrangel Island in the polar circle, and the analysis of fallout particles, made by U.S. experts, almost immediately indicated that the Soviet design involved a completely new approach, which meant a compact deliverable bomb. It was half a year before the Americans were able to do the same, and during these several months the USSR had potential nuclear military superiority.

Sakharov's idea, which was used by parallel research groups, was to build a fission-fusion-fission bomb, in which the central atomic explosion and a chemical "trigger" would start the thermonuclear fusion of deuterium and tritium, which in turn would start off the fusion of light metal lithium. These lithium particles had been found in the fallout.

Soviet books do not mention all these technical details, or Sakharov's role in the history of nuclear development, because Sakharov started to protest against nuclear explosions in 1958, long before his open criticism of the whole Soviet system. However, in some foreign works about Soviet physics, especially in books written in cooperation with Soviet information services to boost the prestige of Soviet science abroad, one can find more details about the real situation (32).

The Soviet success in rocketry technology, which determined the later breakthrough in space research, was also closely related to the brutal and merciless methods of Stalin's time. S. P. Korolev, and many other rocketry and aircraft experts, had been released from imprisonment and rewarded in the very last stages of the war or immediately thereafter. However, they in turn became dictators of large prison-camp systems created to develop military missiles and supersonic aircraft.

The most outstanding German designer of military rockets, Werner von Braun, together with most of his assistants, was already in the United States. However, there were in the Soviet part of occupied Germany about one hundred fully or partially assembled V-2 military rockets, as well as the plants for their production, and a number of technical projects which were still in the form of blueprints (including a two-stage rocket [A-9/A-10] which the Germans were attempting to develop for a transatlantic shot). In addition to all this German technology, about 6,000 German technicians and their families were loaded into trains and sent to the USSR to work in the special prison research centers. They certainly lived better than most Soviet prisoners, but their contribution to the Russian rocketry and space programs was never mentioned in any Russian book about the history of Soviet success in space. Some details about this German technical "help" could be found in Western sources only (see, for example, 33).

Although the dictatorial power given to talented and brilliant scientists helped Russia to get the leading position in nuclear physics, the situation in theoretical physics and mathematics was not so bright. Some physicist-philosophers took leading positions, and they quickly declared the theories of quantum physics, theories of Einstein and many others, "idealistic" and "reactionary." Cybernetics also became a "bourgeois pseudo-science" and was simply forbidden, not only as a research field but even as a subject for study and discussion. This situation later delayed the development of electronics and the computer industry. The gap in computers and electronic equipment became so serious that it has still not been eliminated.

Soviet computer technology lags behind that of the United

States probably by two computer generations. This kind of gap developed also in chemistry, especially in organic synthetic chemistry, where all new foreign ideas had been declared "idealistic." Soviet chemistry had been turned back to the principles of the Russian chemist A. M. Butlerov, a good scientist, but whose main work had been done in the nineteenth century. The economic results of this ignorance became manifest later, in the backward condition of the Russian chemical industry, which was seven to nine years behind the West in the production of many new synthetic materials. Especially disastrous were the consequences of T. D. Lysenko's dictatorial powers in biology and the agricultural sciences. Modern genetics and biology were declared "idealistic," "reactionary," "Morganistic," and a great many pseudo-scientific theoretical and practical recommendations were introduced into education and agriculture.

The privileged position of science and the artificially high standard of living for scientists in comparison with other groups of the population made science a very attractive field for not always talented young people. The special privileges of the scientific leadership led ambitious scientists to develop some new, unknown scientific theories. The old Bolshevik revolutionary Mrs. O. Lepeshinskaya, who had been known as a "good cook" in a small émigré Bolshevik community in Switzerland between 1910 and 1917, but possessed little knowledge of biology and was already eighty years old, announced the creation of a new field of biology which "closed" cellular biology and declared noncellular "living substance" to be the main structural element of all living systems (34). She received official recognition, was elected an academician, won the Stalin Prize along with many other awards, and received much publicity.

The veterinarian G. Boshian "discovered" that viruses could be "transformed" into bacteria, and vice versa. His "discovery" also received publicity as a success in Soviet science (35). Another "innovator," A. J. Titov, declared that he had discovered plants on the planet Mars and established a so-called "cosmo-botany." As the only prominent representative of this new science, he was made a corresponding member of the Academy of Sciences of the USSR and obtained a special laboratory to study life outside earth.

In physiology and medicine the idea of "conditioned reflexes" developed by the late I. P. Pavlov (he died in 1936) became obligatory for the explanation of all physiological processes. Pavlov was a famous scientist, but his research, brilliant and original though it was (he was awarded the Nobel Prize when he was still a young scientist), could not explain everything. However, the idea that the brain had control over all physiological and biochemical processes of human beings became dogma after a special session of the Academy of Medical Sciences of the USSR in 1949. Although the situation was probably not so tragic as in genetics and general biology, it was nevertheless self-destructive. Much of the research on endocrinological regulation and metabolic and other regulatory processes of the physiological and biochemical systems had deteriorated or come to a stop. This, in turn, delayed the development of pharmacology, antibiotics, modern diagnostic methods, and the therapeutic use of endocrinological preparations.

Pavlov's theories had been artificially transferred to psychology and psychiatry, where all modern, "bourgeois" trends were prohibited. (Psychoanalysis as a research method was not known in the USSR until recently.)

Pseudo-scientific approaches became dominant, or at least prominent, in soil science, silviculture, zoology, botany, evolution, agrochemistry, and many other areas. It would be a long and difficult task to discuss all these aberrations and to follow their fate after Stalin's death. Too many of them lived much longer than Stalin, and even Lysenkoism did not die out until recently. Lysenkoism, although its influence diminished, survived as one among many other "research" trends. Lysenko remained in his position as head of a large experimental station not far from Moscow until his death. When he died in November of

1976, his coffin was displayed for mourners in the main hall of the presidium of the Academy of Sciences of the USSR. The memorial meeting attracted some of his old friends and followers, who in their funeral speeches extolled the great contribution of the deceased to the development of Soviet and world science.

The swift development of the scientific network in a country which still faced enormous postwar economic difficulties was not apparent when the prison research centers were built in distant places closed to visitors. In Moscow, however, the symbol of the new policy was evident in the prestigious buildings being constructed on the new site of Moscow University, on the highest hill of Moscow's south suburb. Since 1949 it has been impossible to imagine the Moscow skyline without this university, which is the city's highest and most impressive architectural complex.

Considering the general picture of scientific progress during the period 1945–1953, it is very difficult, however, to find something significant outside the military-oriented areas of nuclear science, rocketry, and aircraft technology. The world's first atomic power station, based on a small reactor, had been under construction in Obninsk since 1952, and it was completed and started to produce electricity in 1954. Obninsk in 1954 was still a "closed" town—a mixture of a prison research center and a "village" of free workers and scientists. When the Soviet press acknowledged the "great success of the peaceful use of atomic energy" (in July, 1954), the location of the power station was not indicated.

There was no shortage of scientific sensations in many other fields of knowledge during the postwar period. Most of them proved later to be fakes—like the "transformation of protein crystals into bacteria" in Boshian's work, the "prolongation of the human life span by sodium carbonate baths," "discovered" by Lepeshinskaya, or the "mutations of rye into wheat, pine into birch tree" claimed by Lysenko. Because Soviet science was almost completely isolated from Western science, criticism from the "hostile" West was greeted with pride rather than concern.

Certainly many branches of science and technology continued to develop slowly, but in some very important fields (genetics, cytology, theoretical chemistry, psychiatry, and others) the movement was mostly backward. Rather than progressing, these research areas had been thrown back almost to the end of the nineteenth century.

The number of scientific institutes and the army of scientists nevertheless grew very quickly. The number of research workers had reached 250,000 at the time of Stalin's death. However, the proliferation of science did not halt the repressions against many scientific schools and groups, not only on the ground of pseudo-ideological principles (as among geneticists) or conflicting theories, but also by the revival of anti-Semitism.

Under the pretext of "Zionist plots," scientists of Jewish origin* began to be harassed, exiled from Moscow and Leningrad to provincial towns, arrested, or dismissed from their positions. This was most evident in biology, physiology, and especially the medical sciences, where the trend culminated in the fabrication of the so-called "Doctors' Plot"—the last of Stalin's wave of terror, which he was not able to complete.† After his death in March of 1953, the turning point in the political climate was felt by the majority of intellectuals when, about a month later, many of the prominent medical professionals and scientists who had been arrested (but were still alive) were released and rehabilitated.

---

* "Nationality" in the Soviet Union does not mean "citizenship," as is usual in many other countries. In all Soviet questionnaires for work or applications for education, and in all passports (until the passport reform of 1976), there were two entries: "Citizenship" (which could be "Soviet," "Hungarian," or "British") and "Nationality" (which could be "Russian," "Ukrainian," "Jew," "Uzbek," "Tadzhik," "Tatar," "Armenian," "Bashkir," and more than a hundred others).

† A group of prominent medical experts, mostly Jewish, who worked or did consultations at the Kremlin hospitals and other special medical facilities had been arrested at the end of 1952 and charged with plotting to poison or to assassinate (through mistreatment) the leaders of the Party and government.

Chapter 6

Khrushchev's Reforms and Scientific Development in the USSR (1953–1964)

The main change during the Khrushchev era was, of course, his bold condemnation of Stalin's terror and the rehabilitation of many millions of victims of political repression.

This rehabilitation came too late for such brilliant figures of Soviet science as academicians N. I. Vavilov, G. D. Karpetchenko, G. A. Nadson, G. K. Meister, N. M. Tulaikov, and many others, who had already died in prison camps. But thousands of other scientists who had been arrested either during the 1930s or during the postwar waves of terror came out of the Gulag Archipelago alive. Some of them were invalids, but the majority were physically and, more important, spiritually able to continue their work. Those who had been released from special prison research institutes were in good shape; their research work had not been interrupted. But many others who did not belong to the technologically or militarily useful professions had spent from five to twenty years in "corrective" camps of hard labor. They were not a "fresh" injection into the scientific community, but they were, nevertheless, extremely valuable and were free from the pseudo-scientific education which had spoiled the younger generation. Quite a few of them were restored to prominence and were able to do important research.

It was not only the technological stars like A. N. Tupolev and S. P. Korolev who, after their rehabilitation,* were responsible for the most striking achievements, such as military and civil TU-type jets and rockets which launched the first satellites and the first spacemen into orbit. There were also some biologists who made their main contributions after long prison or camp terms. A. A. Baev, now vice-president of the Academy of Sciences of the USSR and a Lenin Prize winner for important research in bio-chemistry, spent twelve years in labor camps. V. P. Effroimson, who had been arrested three times for "Morganism" and had spent more than ten years in labor camps, changed his field of research from the genetics of the silkworm to human genetics when he came out of prison at the age of about fifty. He became the most prominent figure in the restoration of the completely de-stroyed field of medical genetics. N. V. Timofeev-Resovsky came out of prison when he was fifty-five, but he was nevertheless still able to organize an efficient research team and to write three books on the problems of evolution, radiobiology, and genetics. Radio-electronist academician A. I. Berg, who was over sixty at the time of his rehabilitation, was the principal figure in pioneering cybernetics research in the USSR, which started in 1955. He established the Scientific Council on Cybernetics and Computers in the Academy of Sciences of the USSR, which became the main supporter of the development of this most important field of modern science and technology. Further examples could be given.

---

* During Stalin's time experts like Korolev, Tupolev, Oserov, and others were *released* as a reward for their job. Their real *rehabilitation*, with elimination of all previous charges, came only after Stalin's death.

The scientific and technological gap between Russia and the Western world which had developed during the Stalin era was not too evident before 1953, because of the almost absolute isolation of Soviet science. Neither correspondence nor exchange of literature with foreign colleagues was permitted. Personal contacts with foreign scientists or scientific trips abroad were practically unknown after 1935. To state publicly or even privately the backwardness of one or another field of research in science or technology was to invite arrest. Popular scientific literature, which tried to attribute all the most important scientific discoveries and technical inventions to Russian scientists, was especially encouraged.

After Khrushchev's "secret" speech at the Twentieth Party Congress in 1956 revealing Stalin's crimes and mistakes in military and industrial leadership, it was no longer possible to hide the real situation. A new picture began to emerge, and the backwardness in science and technology was openly acknowledged; it was possible to put all the blame for this on the late dictator.

The Attempt to Copy Foreign Scientific and Technological Advance

However, Khrushchev as a Party leader of the older generation was not yet able to realize that the division of sciences into "socialist" and "capitalist" was basically wrong. He still thought that the ideological orientation in the sciences was a necessity and was one of the main principles of Marxism. The significant gap between Soviet and Western technology and agriculture was simply explained by Stalin's terror against the best representatives of the Soviet intelligentsia, the informational isolation of Soviet science, and the "class struggle" in the capitalist world, which forced the ruling classes there to invest money in science to develop more sophisticated technical equipment in order to reduce their dependence on the working class and to increase the "white collar" working elite.

Khruschev certainly was most impressed by American technology and agriculture, and though the official propaganda continued to tell the people about the "undefeatable" strength of the Soviet army, U.S. superiority in almost all military technological fields was no secret to Khrushchev and his associates in the Soviet military command. From the very beginning Khrushchev paid special personal attention to the nuclear sciences and the aircraft and rocket industries—the same fields which had received special support from Stalin. This helped Khrushchev to take personal credit for the many advances and striking performances of Soviet technology: first nuclear explosion, first earth satellite, first man in space, first nuclear power station, first shot to the moon, first civil jet airliner (TU-104) on passenger service, and some others. Many of them really impressed the world and even the American general public. But the military balance was certainly not in favor of the USSR. Comparisons of the Soviet and American economies, which became quite legal after Stalin's death, proved how backward the USSR was—sometimes very seriously—in practically all aspects of economic, scientific, and technical development, and the same picture emerged in comparisons with most European countries as well. It was especially damaging for Soviet national prestige that Germany and Japan, whom she had defeated during war, underwent remarkable industrial and agricultural development, and that their growth rate was much higher than that of the Soviet socialist economy.

However, for Khrushchev and the other Party leaders it seemed obvious that the sciences in the Soviet and the Soviet bloc had to be universal and independent of the sciences in the capitalist countries, and that "world" science as such did not exist. Their goal was the same as Stalin's in 1945: to reach the level of science abroad and to surpass it, and to take the lead in all branches of modern science and technology. However, if Stalin's methods for making this possible were connected with putting scientists to work on grandiose projects, such as the All-Union Program for the Transformation of Nature, including changing the climate, eliminating deserts, and constructing gigantic hydroelectric dams and canal systems, Khrushchev's initial steps were simple and seemed logical. His plan was to assimilate the modern methods of American and West European industry and agriculture, to repeat the advances made by foreign technology, and to duplicate "creatively" the technological and scientific work.

The positions of "technical attaché" and "agricultural attaché" were established in many Soviet embassies abroad, delegations of Soviet scientists started to make excursions abroad to study some special branches of industry and science, participation by Soviet scientists in international congresses and conferences, though limited, became possible, and most foreign scientific literature became freely available.

One of the most important initial demonstrations of Khrushchev's new policy in science and technology was the Soviet Union's extensive participation in the First United Nations Conference on the Peaceful Uses of Atomic Energy in Geneva in 1955. Most research in this field had been classified in the USSR. Academician Igor Kurchatov and a special government commission were responsible for the composition of the Soviet delegation. They had two choices: restrict the number of papers and discuss unimportant and inevitably poor works, or declassify a great many works and show the true face of Soviet nuclear science. Khrushchev approved the second choice; he wanted to impress the Americans, and it was a good decision. Of course, not all nuclear research was declassified, but the ice was broken. The monthly journal *Atomnaya Energiya* (Atomic Energy) began commercial publication in the USSR, and the periodicals *Problems of Cybernetics, Space Research,* and others were starting to appear, a development unimaginable during Stalin's time, with his obsession with secrecy.

With some reservations, Khrushchev approved Soviet participation in the Pugwash international conferences, at which prominent scientists discussed world political problems and the social implications of different projects. The First Pugwash Conference took place in 1957, and it was a mixed meeting of scientists and diplomats. The American billionaire, Cyrus Eaton, who financed the movement, established a friendly personal relationship with Khrushchev. The main problems discussed by the Pugwash conferences were usually disarmament, arms control, and international cooperation. These conferences probably did not have much influence on the solution of these problems, but they certainly raised the consciousness of Soviet scientists about their role in their own society and in the world.

One of the advantages of the "duplication" orientation of Soviet science was the creation of a comprehensive system of scientific information. A special Institute of Scientific Information within the Academy of Sciences of the USSR was transformed into the All-Union Institute of Scientific and Technical Information. It received extensive technical facilities enabling it to publish many journals of abstracts which covered world science and technology, to translate special "express" information which institute staff members considered to be of special importance, and to provide photocopying services where scientists could order photocopies of any article from both foreign and Soviet journals. Institutes of Scientific Information had also been created within the Academy of Medical Sciences of the USSR, the Academy of Agricultural Sciences, and in some industrial research networks. The system did not work very quickly or efficiently at the outset, but it very soon acquired the necessary experience. It was staffed by qualified scientists and editors, and thousands of scientists from other institutes helped in the preparation of journals of abstracts (in chemistry, biology, electronics, etc.).

The publishing house Mir, set up after the war to translate foreign literature, was also substantially expanded and initiated a new and very extensive program of translations of the most important works of foreign scientific literature. These developments meant a sharp increase in the amount of information available about foreign scientific research.

The All-Union Institute of Scientific Information became responsible for the photo-duplication of original foreign journals for internal subscription, a procedure which I criticized on the grounds that it caused delay and because fresh and original research information was often censored (9). When I criticized this procedure, in 1968, it was certainly already obsolete and unable to satisfy the demands of advanced research. But when it was first introduced (in 1953–1955) and developed (in 1956–1958), it was a significant stimulus to science after dull years of postwar isolation. Only a small minority of scientists could read foreign literature in the original languages in 1953. Before the war, German was the foreign language most generally taught in secondary schools and universities, but after the war German lost its importance as a language of science. Therefore, comprehensive translation facilities and abstracting services were very important for Soviet research workers. In 1967–1968, English had already become well established at all levels of the educational system, and at least a good reading knowledge became obligatory for postgraduates.

Many foreign exhibitions were organized in the USSR, and foreign currency became available for purchases abroad of examples of different kinds of equipment from different countries. The Soviet Union was not a member of the Patent Convention and was free to duplicate any kind of foreign design without purchase of license. This seemed to be an advantage for the time being. The duplication ("assimilative repetition") trend in research and technology was inevitably related to the proliferation of research and testing systems and the hypertrophy of scientific establishments. It was the easiest time ever for one or another more or less prominent scientist to propose a big institute or a laboratory. Sometimes it was enough to prepare a detailed report for the Party Central Committee or the government that such-and-such very important branch of science or technology was well developed in the United States but very badly represented in the USSR.

It was not only the United States which was cited as an example. The All-Union Research Institute of Gerontology in Kiev was established after a few articles in newspapers had declared that scientists in Romania (Anna Aslan among them) had found methods of prolonging the human life span by Novocain* injections (36). The publicity given in the Soviet press to the Romanian findings induced the Soviet Ministry of Health to start the program of Novocain therapy in the new Institute of Gerontology. The first few years of the institute were spent testing this prolongevity method, but after several years of research the method was dropped as the result of negative conclusions drawn from its rather wide use. (The method was tested extensively in almost every developed country without statistically reliable positive results, but it is still in practical use in the United States and elsewhere.) The institute, nevertheless, survived.

As a gerontologist myself, I was partly involved in the discussion about the failure to repeat in the USSR the Romanian method of prolonging the life span by Novocain. An explanation for the difference in the efficacy of the method in Romania and the Soviet Union was discovered, but never published, since it was rather delicate from the political point of view. In Romania the Aslan clinic procured most of its patients from old-age institutions, where conditions and medical care for aged and often homeless people left much to be desired. Once placed in a first-rate government clinic, these patients responded very positively to procaine injections combined with vitamins and good care. In the Soviet Union the first people to undergo revitalization therapy were top-ranking government and Party officials, including Nikita Khrushchev himself. Procaine injections produced no objective revitalization effect in them (only temporary euphoria).

One of the products of this research boom was the Research Institute of Medical Radiology in Obninsk, where I started to work in 1962. This institute was established in 1959, after the Second United Nations Conference on the Peaceful Uses of Atomic Energy. The backward position of Russian radiology and

* Also known as procaine, Gerovital, and Aslovital.

radiobiology was exposed at this conference and led to the creation of several institutes of radiology and radiobiology in different parts of the Soviet Union.* The institute in Obninsk was
the biggest project among them. It also had a special technological duplication task: to buy examples of radiological and
roentgenological (X-ray) equipment from different countries
(Germany, the United States, Japan, Britain, Italy) and to develop "original" Soviet models of such equipment by the assimilation of the best from different examples. A large workshop,
which would be able to fabricate "Soviet" advanced models of
such equipment, was an integral part of the institute, and the
projected staff of the institute was about 2,000, a figure which
was, later, easily overfulfilled. Even this large staff was not
enough for all the experimental and clinical divisions of the institute, which filled almost twenty buildings and was under construction for about ten years. But the neighboring Research Institutes of Radiochemistry and Nuclear Energy were much
bigger; they had about 3,000 staff members and 6,000 workers,
as well as extensive facilities for duplicating research.

The idea of assimilating foreign technology and science was
not a bad one, if it had been supplemented by real cooperation in
science and the integration of research on a world scale. But the
ideological barriers to integration were still too strong. However,
this duplication was based on the wrong assumption that scientific progress in the capitalist world was a temporary phenomenon of postwar reconstruction that would inevitably be halted by
the industrial crisis of overproduction, which was only delayed by
the boom in the postwar economy. Khrushchev and his Party colleagues could not imagine that scientific and technological prog-

* The establishment of several new research institutes of radiology was also
stimulated by the increase in radiation sickness among people employed in
the nuclear and radiochemical industry or personnel working with radioisotopes. New safety regulations and strict new rules for work related to radiation
and isotopes, which had been introduced in 1958, reversed the rapid growth rate
of this hazard.

ress in the West could last, and on this assumption they recommended the areas in which leading positions were to be reached. Unexpectedly, though, scientific and technological progress (called the "Scientific-Technical Revolution") did not come to a halt, but accelerated in both the United States and Western Europe in the late 1950s and 1960s. Moreover, the duplication of modern sophisticated equipment for science and industry was a rather slow process. In my book on international scientific cooperation (9)* I gave several examples of how duplication did not decrease but increased the technological gap, because in the time it took to duplicate some pieces of equipment, foreign technology was able to introduce so many changes that the new equipment was several generations ahead of the Soviet "model" when it finally became available for industrial production, and it was already obsolete. This was the case in very many areas. The Khrushchev policy also meant an enormous increase in the number of research institutes and research staff. The number of research workers had almost tripled since 1953 and reached 650,000 in 1964. The number of research units (of research institute type) reached the figure of 4,800—almost double that in 1953. The number of engineers graduated from higher technical schools and universities, which was 37,000 in the USSR and 61,000 in the United States in 1950, showed a dramatic reversal by 1965 (41,000 in the United States and 170,000 in the USSR [37]).

However, this impressive growth of the scientific and technological base did not reflect the real economic and agricultural development of the USSR. Soviet economy and agriculture have lagged behind the United States and Western Europe increasingly since 1958. The agricultural troubles, which were most closely felt by the population because of the serious shortage of food, developed into the crisis of 1963, when the Soviet Union

---

* This book was published abroad in 1971 but I wrote it in 1966–1968, when the "duplication" trend was still very strong.

tried to avert famine by enormous purchases of grain and other foodstuffs from abroad. Of course, it was not science that was responsible for this crisis; the background of Khrushchev's economic failures has been described by Roy Medvedev and myself elsewhere (38). Khrushchev's main economic reforms—the abolition of all state-controlled machine-tractor stations in the country within one year, end of centralization in industrial management, attempt to accelerate meat production beyond a reasonable rate, promotion of corn planting in climatically unfavorable areas—proved to have negative consequences in the long run. His attempt to divide the Party system into industrial and agricultural sections (instead of a territorial network) and to introduce compulsory turnover into the Party apparatus at all levels induced dissatisfaction among the Party bureaucracy. The decision to reduce the army made inevitable the confrontation between Khrushchev and the military establishment. There were many other failures and political reasons which made inevitable Khrushchev's forced resignation in October, 1964. But for a better understanding of the subsequent trends in the development of Soviet science, it is necessary to consider some special reforms which Khrushchev made and which were irreversible.

The Compulsory Merger of the Applied Sciences with Industry and Agriculture

The privileged position of science and scientists in the USSR varied among different fields of research, in different administrative systems of the scientific establishment, and in the different locations of scientific research institutes and universities and higher schools. After an initial substantial increase in state investments in science in 1945–1946, the rapid growth of research facilities and research workers made it necessary to reduce, in some way, the rate of budget growth from year to year, because the general economic growth was not as substantial as the growth in science and education. In 1951 a new regimentation divided all scientific and educational units into three main categories. Research institutions of the first category were those judged to be of special importance, and these first-category institutes had priority in receiving the imported equipment, acquiring new buildings, filling staff vacancies, and being paid according to a higher scale of salaries for all groups of research scientists. The second category of research establishment took second position in all these privileges. The institutes, higher schools, and experimental stations of the third category had reduced supplies, reduced attention from state officials, and the lowest salary scale in the world of Russian science. I do not plan to compare all the categories in all aspects, but will merely indicate that the middle research worker, a comparatively young scientist with five to six years of experience and with a first scientific degree of "candidate of science" (somewhere between a master's and a Ph.D degree in the British scale), working with a small group but without administrative obligations, would receive 3,000 rubles per month (or 300 rubles after the currency reform of 1961)* in an institute of the first group, 2,500 (250) rubles in an institute of the second group, and 2,200 (220) rubles per month in a research establishment of the third category. The same ratios (100:83:73) were valid for all other positions, such as heads of laboratories, junior scientists, professors, and even directors of the institutes.

For understandable reasons all research scientists, if they were given a free choice, would consider themselves first category; hence the decree about this regimentation was accompanied by a list of all research units already divided into these groups. The Academy of Sciences of the USSR, as the top scientific establishment, was considered a research group of the first category, and all research institutes dealing with military and classified work

---

* Currency reforms in 1961 introduced a new, "heavier" ruble: one new was equal to ten old rubles. At the same time, the ruble was devalued against foreign currency. One dollar is equal to .07 rubles now, but the rate in 1961 was about 1 to 1.

were also in this first priority group. The institutes of the Academy of Medical Sciences and the Academy of Agricultural Sciences, as well as the institutes of academies in the provincial Soviet Republic, were divided between the first and the second groups. Those research institutes which were in Moscow, Leningrad, Kiev, and other big cities and which already had very high reputations were listed as in the first group; those situated in smaller towns and which were lesser known became members of the second group. The rest of the research institutes and the laboratories or experimental stations which were not united within different academies' networks or which were not connected with classified research were put into the third category. These were the experimental stations in agriculture and the institutes connected not with the academies but with the ministries and concerned with various local industrial and agricultural problems. Sometimes the differentiation into categories was within the same institute or higher school. For example, two research units had been attached to the Department of Agrochemistry and Biochemistry in the Timiriazev Agricultural Academy in Moscow, where I worked between 1951 and 1962. One, an experimental station with a biochemical laboratory, where I was a senior research scientist, was in the third category, whereas the other, the Laboratory of Biophysics, was in the first category because it carried out some research in radiobiology which was classified.

This differentiation, probably justified from the point of view of the need to reduce the growth in the science budget, later had a special influence on the development of Soviet science. The differentiation applied to existing institutes and units, so when a new institute was established, the government had to decide about its rank in this three-level system of priorities and salary scales. Since all branches of the Academy of Sciences of the USSR were within the first group, all new institutes in the Academy also had to be in the first class. In other systems budget difficulties usually made such immediate recognition difficult, and this was especially true of provincial cities and towns. In such a situation many ministries and industries which wanted to establish new research institutes with a clearly practical industrial or agricultural orientation nevertheless approached the Academy of Sciences of the USSR with the suggestion of establishing the new institute within the Academy system, with a promise to finance the research activity and staff salaries of the institute. If the industrial institute—for example, the Research Institute of Oil Industry, Institute of Coal Mining, Institute of Polymer Chemistry, or Institute of Metal Technology—was established within the Academy of Sciences of the USSR and located in Moscow, Leningrad, Kiev, or some other attractive large city, it was put into the first priority group, and both its central location and its high salaries allowed it to employ better and brighter scientists of all ranks. The director of such an institute would eventually receive the title of "academician," and the whole unit would have more prestige than would be the case if it were more closely related to its branch of industry, and more reasonably situated in the industrial centers producing oil, coal, iron ore, steel, or railroad equipment.

The Academy of Sciences, which was before the war mainly the center of the pure sciences—physics, biology, chemistry, and mathematics—and to a lesser extent the humanities, began to grow into an enormous organization with complex connections with all branches of industry and agriculture and with a clearly dominant role in the applied sciences. The new building areas in the southern suburbs of Moscow were saturated by dozens of new research institute buildings—all, of course, in the first priority group, reflected even in their expensive architectural designs.

Khrushchev did not like this "industrial" growth of the Academy. The high concentration of institutes of applied sciences in Moscow, far from their respective industries, prevented them from having much influence on the technical reconstruction he had tried to bring about in so short a time. His appeals in speeches to the Academy about changes in the Academy structure had no effect, nor did the special government ban on the proliferation of new administrative buildings in Moscow. The program of residential development included the building of primary and secondary schools for children. Hence the Academy of Sciences of the USSR, with financial help from many ministries, simply started financing the Ministry of General Education in the construction of many new school buildings, with a view of letting the empty ones for new research institutes. The Ministry of General Education, which was under no restriction in the building of new schools, happily accepted the assistance of the Academy, and would construct five, six, or ten new school buildings in poorly populated areas of Moscow whose local population had not enough children to fill them, so they had to be either sold or let to the Academy of Sciences of the USSR. Not only technical institutes, like the Institute of Electronics, got Moscow residence in this way, but many theoretical research institutes, like the Institute of Developmental Biology, acquired secondary school buildings, which they converted into laboratories and other facilities according to their needs.

Finally the situation reached an intolerable level, and Khrushchev's difficulties with the Academy were discussed at the Party plenum in April of 1961. The decision of the Central Committee and the government reformed the scientific structure of the Academy with one blow; about half of the research institutes of the Academy were distributed among different industries and were relocated in areas outside Moscow, Leningrad, and the other big cities. The Academy of Sciences of the USSR was restricted to developing mainly the theoretical sciences. However, the institutes that were relocated in the new areas did not in many cases contribute much to industrial development. Residence in Moscow, Leningrad, Kiev, and Minsk had such advantages for scientists that many of them, especially the more prominent ones, did not move to the new, rather less attractive locations. This was also true of technical staff and qualified workers. In the new locations neither good buildings nor qualified staff were readily available. So the too hastily improvised reform in fact destroyed some important branches of science, and it took many years to reestablish facilities and staff able to carry out serious research work. Some of the new locations were Krasnoyarsk (Siberia), for the Institute of Forestry; Ufa (an oil production center), for the Institute of Oil; and Kaliningrad (former Königsberg of East Prussia) for the Institute of Fisheries.

The Academy of Sciences of the USSR was reduced to about half its size, but the loss of research personnel was not so dramatic (about 25 to 30 percent [1]), because many scientists preferred to stay in Moscow and to take less responsible jobs, but within the Academy system. Former heads of laboratories of institutes that were to be relocated could agree to stay in Moscow as junior scientists, and, of course, very few academicians moved into the remote areas. The reform would not have disrupted the applied sciences to such a degree if the relocation had been earlier and more gradual, or if all those hundred or so industrial research institutes which, against all common sense, were established in Moscow in 1950–1960 had from the very beginning been situated in their natural locations.

The Decentralization of Pure and Theoretical Sciences

Khrushchev's other major reform in Soviet science was related partly to his general program of decentralization and partly to the impressions he had received on his many foreign trips. He liked the small research centers which exist outside the large industrial cities in the United States, Great Britain, France, and other countries. He therefore decided that the pure sciences would be better developed if research workers lived and worked in quiet, attractive places. The concentration of scientific institutes in Moscow and a few other big cities during Stalin's time had reached too high a level. In biology more than 70–80 percent of serious scientific research was carried on in Moscow, and 90 percent of the research journals in all branches of science were published in Moscow.

Such a concentration of science in one place was considered strategically dangerous, and discussions about decentralization became unavoidable as soon as the discussions about the decentralization of industrial management and the liquidation of centralized ministries were launched in 1957. The industrial and agricultural development of the sparsely populated but potentially rich eastern parts of the Soviet Union (West and East Siberia, the Far East, Kazakhstan, etc.) had been an important point of the Party program since Lenin's time. (In fact, this had also been the main direction of imperial Russian expansion since the reign of Ivan the Terrible, who joined the Siberian territories to Russia in the sixteenth century.) For the more rapid development of these areas, Khrushchev felt it necessary to establish modern research centers there.

Many new, prestigious scientific projects in physics, nuclear physics, astronomy, geophysics, space research, chemistry, or even in plant physiology or farm animal breeding, could not be undertaken in cities; they needed large construction sites that were not available in urban areas. All this stimulated a healthy, but, again, too hastily developed, program of decentralization of theoretical and pure science and the creation of special scientific small towns located usually in attractive country places. The first models of such scientific towns were established by academician Igor Kurchatov during Stalin's time—for example, the now famous town of Dubna on the Volga River, about eighty miles north of Moscow, which is a big center for theoretical research in atomic physics.

But the towns built during Stalin's time were usually closed centers for secret research. Khrushchev, who was not so obsessed with secrecy, made Dubna an open town and transformed it later into the International Center of Research on the Peaceful Use of Nuclear Energy. I have already mentioned Obninsk; it has a similar history, but Obninsk was a closed town for a longer time, and when it was finally opened, it was nevertheless closed to foreigners.

The small scientific towns around Moscow were satellite towns, usually established when it was necessary to build a research complex with nuclear reactors, or when large areas for radio telescopes and other equipment for space-oriented work were required. The same applied to agricultural and geophysical research, which required experimental fields and a quiet environment. Much technical research had also been transferred out of Moscow or Leningrad, because it too was unsuited to urban conditions—the testing of new jet engines, for example, which was too noisy for a large city. These special conditions gave birth to not less than a dozen small satellite research towns situated at a distance of thirty to seventy miles from Moscow in all directions along the railroads or main motorways.

The creation of big new scientific centers very far from Moscow was quite a different development pushed through by Khrushchev. The best known and most important among these special projects was Academgorodok (Academy Town) near Novosibirsk, the main West Siberian industrial center. The big hydropower project, started here during Stalin's time on the upper part of a major Siberian river, made a large artificial lake. The shore of this lake was selected for a new research center. To give the center prestige, it was made a special Siberian division of the Academy of Sciences of the USSR, and several well-known academicians (among the pioneers were the physicist M. A. Lavrentiev, the mathematician and cybernetician S. A. Sobolev, and the geologist S. A. Christianovich) decided to move there to establish new research institutes. The governmental decision of May, 1957, made available unlimited support for the building of the new center, and when I visited the place in 1960, three years later, it was already a town with a population of about 60,000 and had several newly working institutes and a new university.

At its very beginning the center started to build fourteen new research institutes. The most important were the Institute of Nuclear Physics, the Institute of Catalysis and Chemistry, the Institute of Organic Chemistry, the Institute of Geology, the Institute of Applied Mathematics, the Institute of Hydrodynamics, and the Institute of Cytology and Genetics. The initial project was quickly overfulfilled, and many new research institutes were added. The eastward movement of theoretical science played an important part in the general program of the decentralization of science and the economy. Not only could research into local resources be carried out more effectively, but the establishment of scientific facilities in the eastern part of Russia created better conditions for the industrial development of this area.

However, Khrushchev, as he often did, made one major mistake, which almost nullified the results of the scientific development of the Siberian part of Russia. Working conditions in Siberia have always been considered especially difficult. Therefore, the salary of workers in all industrial places east of the Urals had a differential northern increase, sometimes 50 percent, sometimes double the salary for the same work in the European part of the USSR. Khrushchev's programs always strained the budget, and he therefore constantly tried to find new ways of financing his projects, within the USSR as well as abroad. The scientific boom in Siberia gave the impression that conditions were not too rigorous, and the thousands of scientists who moved there gave places which until only recently had been associated with dozens of prison labor camps some aura of respectability. This new respectability, reflected in all propaganda, made it possible for Khrushchev to cancel unexpectedly the differential payments for workers there, and to make salaries more uniform throughout the country. This meant a reduction of income for many millions of Siberian families. The measure, introduced in 1960, did not produce any open conflict; many former camp prisoners and exiles still lived there, not having found it possible to move to a better climate immediately after release, and finally having settled around the former camps. These people were probably afraid of any open dissent against the new economic measures. After 1960, the migration of workers was westward, so although science had moved east, the industrial power of the new areas did not grow. The population censuses of 1959 and 1970 revealed a decrease in Siberian population, while in the European part of Russia and in all southern areas of the USSR population growth was rather significant. But that is another story.

Khrushchev's Reforms in the System of Promotion and Qualification

While considering the problems of the development of science and the scientific community in the USSR, it is necessary to identify who can really be qualified by the term "scientist." "Science" in Russia was always a term with a rather broad meaning, probably equivalent to *Wissenschaft* in German. All kinds of research work was considered to be scientific, and therefore scientific degrees and academic titles could be given to representatives of practically all intellectual professions. Not only biologists, physicists, and chemists could be "candidates of science," "doctors of science," "docents," "professors," "academicians," or "corresponding members of the Academy," but all these degrees and titles could also be bestowed on technologists, engineers, lawyers, literary critics, philologists (linguists), philosophers, military experts and so on. However, there are some differences between scientific degrees and academic titles. The degrees (*stepen'*) "candidate" and "doctor" must be defended in open debate at the academic council of some scientific institution (a prestigious research institute, a university faculty, etc.) on the basis of a thesis prepared after some years of research work, whereas a title (*zvanie*), such as "professor," "docent," "corresponding member of the academy," or "academician," is awarded without a thesis as a result of promotion confirmed by a secret ballot of the academic council or (for the academies) by a general meeting of the academy members. No open defense of a thesis is necessary to receive the title; however, the discussion about the qualities and standing of applicants for titles was and is obligatory.

There are many reasons for young scientists to try to obtain higher degrees and titles, but the main reason is that the salary scale is geared to them. In a research institute of the first category, for example, a senior research scientist without a degree will have a salary of about 120 rubles per month; * with a "candidate" degree his salary will be more than doubled—300 rubles per month (about $400, or £220, according to the 1977 rate of exchange); with the degree of "doctor," 400 rubles per month; with the additional title of "professor," 500 rubles per month; with the title of "corresponding member of the Academy," 750 rubles per month; and with the title of "full academician," 900 rubles per month—all differences without a change in position. † The same variations, on a different scale, apply to positions within research establishments of the second and third categories, and for universities and higher educational units as well.

But general figures for the number of scientists in the USSR will apply to all research workers with a higher education, both with and without scientific degrees. Postgraduates (aspirants) working to prepare their theses are also considered research workers.

* This is about the same as the income of an industrial worker. Although it seems very modest by American standards, one needs to take into consideration that not more than 10 percent of the salary goes toward rent, electricity, heating, etc. Medical service is free, the cost of transportation is very low, and education at all levels is also free. The maximum rate of income tax for higher salaries is 13 percent.

† Salary increases are decided twice, after five years' and ten years' work, and do not change later. In both cases the increases are fixed at 10 percent. Because inflation in the USSR is almost invisible and many prices have not changed since 1951, salaries for research workers are more or less stable, and the figures I have given here for 1951–1956 are still the same for scientists with degrees and titles. The low-income groups (research workers without degrees or titles) did have small increases, however, of 10 to 15 percent between 1955 and 1975. For research workers in medical fields, the salary increase of low-income groups was more substantial.

Without having a scientific degree, received after open defense of a thesis, one cannot do really independent research, nor can one obtain a senior position with the academy or university networks. The first degree which opens promotion possibilities is "candidate of science," which is approximately equal to the Ph.D.* In 1976 in the USSR there were 375,700 research workers with the degree of "candidate" and only 34,600 with the degree of "doctor" (the corresponding figures in 1950 were 45,500 and 8,300), but the total number of research workers was 1,254,000 in 1976 and 162,000 in 1950 (66).

The prestige and high standard of living associated with research work starts with the first scientific degree of "candidate of science," which officially opens the possibility of competing for senior positions—senior research worker, head of a laboratory, docent, assistant professor, member (or corresponding member) of specialized, republic, or even central academies of science.

The usual method of obtaining this degree would be by undertaking original research and defending it before a respected group of scientists who are members of the "academic council" of one or another research institution. The list of such councils which have the right to give the degree is approved by special decision of the Council of Ministers. Usually it is necessary that no fewer than eight or nine doctors of science be members of such a council, and sometimes members are from different institutes. (In 1958 there were 548 academic councils in the USSR with the power to consider and award scientific degrees. At that time the Highest Qualification Commission rejected 88 doctoral and 252 candidate theses as inappropriate.)

The usual time required for a paid postgraduate (aspirant) to prepare a thesis is three years, but a junior scientist and even a technician with higher education could prepare a thesis for de-

---

* Sometimes "candidate of science" is translated "master of science," but this is a wrong comparison. "Candidate" theses and examinations are about equivalent to those for the Ph.D., but the whole system for awarding the "candidate" degree is much more complex and multistaged.

fense while continuing salaried work in different positions. (In addition to defending a thesis, several professional examinations and one compulsory one on Marxist-Leninist philosophy are required for all professions as well.) The defense of a thesis is a special case. Each thesis must be read and evaluated openly by two "official opponents," who give their opinion in written form. Members of the council can take the floor for debate, and all who are present at the open defense can express their opinion about the defendant. The rules adopted after 1947 state that a synopsis of the thesis (ten to twenty pages) must be printed and distributed among the research institutes, universities, and laboratories of the same type (physiology, biochemistry, etc.) and sent to the scientific libraries; scientists from other places who want to criticize the work can send their reviews and opinions to the secretary of the council to be read during the defense proceedings.

So the whole process of obtaining the first candidate of science degree is quite long and time-consuming. But it is even longer for those who want doctor of science degrees. The doctoral degree can be awarded only to those who have already received the candidate degree, but a new thesis must be written and defended openly to obtain it.* The number of scientific councils with the right to give doctorates are fewer, and these are usually the councils of organizations of very high prestige. The manuscripts of doctoral theses are longer (usually they are monographs of 500 or 600 pages), three official opponents are necessary for the de-

---

* The full reading of the doctor of science degree is not just "Dr."; it is obigatory to indicate the area of scientific specialization. Therefore, it is normal to write *before* the name of the scientist "Doctor of Biological Sciences," "Doctor of Physico-Mathematical Sciences," "Doctor of Medical Sciences," "Doctor of Technical Sciences," "Doctor of Law Sciences," "Doctor of Pedagogical Sciences," "Doctor of Agricultural Sciences," "Doctor of Philosophical Sciences," and so on—fifteen or sixteen kinds of degrees. There are cases of double degrees: one person could be at the same time a "Doctor of Biological Sciences" and a "Doctor of Medical Sciences," and both degrees would be indicated before his name. The same is true of candidates ("Candidate of Biological Sciences," "Candidate of Historical Sciences," etc.).

fense, and then the degree as such is only recommended. The final decision is made by the Highest Central Attestation Commission for Scientific Degrees and Titles at the Ministry of Higher Education after the thesis has been considered by a special board of experts and both the board and the plenum of the commission have voted by secret ballot. In 1958 the Central Commission board of experts controlled all candidate degrees as well, but this was only control, not the final decision. So it took four to six months after defense to get the diploma for candidate of science, but one or two years could elapse before the doctor of science degree was final.

It is important to indicate that before 1940 the Academy of Sciences of the USSR and the specialized academies had the right to give both candidate and doctoral degrees to scientists without the process of open defense, simply by the decision of the academy presidium, if the group of academicians or the research institute of the Academy nominated somebody for such qualification. Since 1947 the academies of sciences have been stripped of the power to award scientific degrees without open defense and the writing of a thesis.

The road to a scientific degree is, as we have seen, by no means simple, and the whole system was always under attack because of its complexity. Many scientists expressed the view that the writing of long theses just for the sake of obtaining a degree was a wasteful process, and that aspirants to candidate and doctoral degrees must do something in addition to their current research more noteworthy than writing a thesis, even if its synopsis was published and distributed among one hundred to two hundred laboratories and departments.

But Khrushchev's approach toward scientists was different. He thought that too many of them were seeking scientific degrees just for an increase in salary, and that this was why they did not always work productively enough to make a real contribution to science and technology (which, of course, was often the case). However, he was sure that scientists would work more productively if the road to the privileged position associated with a scientific degree had certain conditions attached to it. The system of qualification, which was already complex enough, was made even more complex and multistaged in 1956. The Party Central Committee and the government drafted a decree "About Improvement of Scientific Qualification" which introduced new elements into the whole system. To be permitted to openly defend his thesis, the applicant had to get the main part of his research published in a scientific journal. Theses could be offered for defense three or four months after such publication. This was a serious obstacle, because Soviet research journals had large backlogs and the average waiting time for publication was often more than a year. But this was not enough. To make sure that science was more practically oriented, the new regulation also made it obligatory that the thesis must be evaluated not only by the two "official opponents" named by the council (or three in the case of a doctorate) but by one or another industrial, agricultural, medical, or other organization, which could judge the practical value of the defendant's recommendations. To make the consideration of theses more independent, the new regulations forbade the defense proceeding to take place in the scientist's own research institute or faculty. This meant that a thesis prepared, say, in Moscow University must be defended in other universities (Kiev, Leningrad, etc.) with Moscow University excluded. For reasons of economy, the increase of salary now started not after the decision of the academic council, as formerly, but only after the valid diploma had been received from the Highest Commission at the Ministry of Education.

The new regulations had very little effect on the quality or the practical orientation of research related to the preparation of a thesis. But the immediate general effect on the number of degrees and titles awarded was negative. In all branches of science the annual "production of scientific staff of high qualification" dropped below the 1947 level. When the number of officially qualified scientific personnel with candidate or doctoral degrees fell far below what was necessary to fill vacancies in quickly growing research networks, the strict new regulations concerning "prepublication" and "practical review" and the ban on defense before one's own academic council were changed, and the growth rate of "highly qualified scientists" accelerated.

However, the obsolete system of double defense was not changed, and it is still in force today in spite of being contrary to the normal age distribution for scientific productivity. Most recipients of the doctoral degree (which is very important for many responsible positions in the scientific hierarchy) are between fifty and sixty years of age (36 percent), with as many as 15 percent receiving this degree when they are over sixty (39). 37 percent are awarded the degree between the ages of forty and fifty, and only 14 percent between the ages of thirty and forty. Research workers younger than thirty who are awarded this degree are so rare that they are not even considered by statistics (three to seven cases per year [39]). There are quite a few cases of scientists' completing their theses for open defense when they are sixty-five or seventy. I once was present at the defense of a thesis for a Doctor of Biological Sciences degree by a biochemist who was eighty-one years old. The stress of the whole procedure, together with the subsequent ordeals of "expertise," secret ballots by academic councils, and the plenary session of the Highest Commission, is so great that the incidence of heart attacks and general mortality at the doctoral stage of a scientist's life is much higher than at any other period. Those who have passed through all the stages of scientific qualification, while they may not necessarily be good scientists, must be strong human beings. Probably this explains why, in spite of the fact that the average age of a full academician of the Academy of Sciences of the USSR is about seventy, academicians in general have a better chance than the average person of living to ninety or more.

The Breakthrough in Space Research and Its Influence on the Development of World Science

As we have seen, Soviet science and industry were able to carry out the nuclear program in a speedy and impressive fashion. The design of the hydrogen bomb in 1953 and the construction of the first atomic power station in Obninsk in 1954 were obvious landmarks which diminished any inferiority complex in the area of technical competence. Although both the testing of the hydrogen bomb and the commissioning of the first small atomic power station took place after Stalin's death in March, 1953, these achievements were the final results of Stalin's postwar directives regarding the main objectives for science. The building of the atomic power station did not attract much attention outside the USSR, because it was a small experimental station whose experimental attachments probably consumed more power than it produced (5,000 kw.). This station tested a small-size reactor which had some practical usefulness for atomic power units later built for nuclear-powered ships and submarines. Similar small reactors were also available in the United States, and some nuclear-powered submarines had been tested or were under construction. The Soviet H-bomb had a more serious impact abroad, but in the context of the nuclear arms race rather than of general scientific development.

The first real sensation almost entirely attributable to Khrushchev's new policy and to the enthusiastic and genuinely independent research and engineering work of Korolev's research unit was the launching of the first man-made earth satellite (*Sputnik*) on October 4, 1957. Secrecy increased the impact of the event, because it had all the advantages of surprise. The episode seems even more remarkable now, when the story of the Soviet space program is well known and has been described by many Soviet and Western authors. Soviet science and technology in the field of rocketry was, in 1957, well below the U.S. level. The Americans had also carried out very costly research in preparation for launching a satellite. Their program was not as secret, and the popular and specialized press published information about satellite design, the electronic and photographic equipment for a satellite, its weight, the type of special multistage rocket available, and even the possible date of the actual launch. At this stage Soviet designers had neither a rocket which was powerful enough nor the sophisticated electronics to match the American space program.

But Korolev, who had been a space research enthusiast since the early 1930s, well knew that in space science, as in any other branch of science, the real discovery belongs to the *first* to make it, even if it is with primitive equipment, and not to the *second,* who merely repeats it, even if at a much higher technical level.

Korolev designed his intercontinental rocket, able to carry a nuclear weapon, in 1955. It was tested more or less successfully in 1956 and was accepted by the army. This design was probably expensive, but it represented a very simple solution to the problem. It had the nickname *semiorka,* or "septet." It was a cluster of seven small rockets together with a specially synchronized booster. The cluster system was much heavier than the single American rocket designed for the same purpose, but it worked, and it could carry more weight in the second and third stages. It could easily be transformed into a multistage rocket with the final stage accelerated to the first escape velocity. Korolev presented his project for the first "surprise" satellite for Khrushchev's approval at the beginning of the summer of 1957. It was approved within a few weeks. Khrushchev was always obsessed with the idea of demonstrating the superiority of the Soviet system over the American in something, and just at this time he narrowly escaped defeat as a result of an "anti-Party" plot when Molotov, Malenkov, and Kaganovich organized a strong anti-Khrushchev group in the Presidium of the Party's Central Committee, using as a pretext Khrushchev's adventurist statement made in Leningrad in May, 1957, when he declared that the USSR would catch up and overtake America in the production of meat, milk, and butter within three years. This promise was utopian and subsequently proved to be unrealistic, but the idea of demonstrating Soviet ability to surpass the United States in the technological area, where the absolute superiority of America was universally acknowledged, had enormous propaganda value for Khrushchev.

A successful test of the new rocket was carried out in August, 1957. Korolev was a rocket expert, and therefore there was at this stage no real technical development of the satellite itself. The last stage of the rocket was heavy enough to carry a lot of different equipment. But to design and test such equipment (which was the main part of the American space program) needs time, special research units, and a great deal else. Time was short, however, and Korolev realized that a satellite is a satellite, even if it is an empty sphere with a powerful shortwave transmitter broadcasting simple signals.

The satellite was launched on October 4, 1957. Its signals from space had a more sensational impact than anybody could have foreseen. American and Western technological and military superiority were so obviously challenged, and the general public so impressed, that all channels of financial support for science and technology in the West were opened wide. The cold war was not yet over in 1957, and the fear that "the Russians are coming" approached panic in some quarters. Financial and general support were not restricted to the space program; practically all fields of research in the West—chemistry, biology, medicine, and so on—benefited from an unprecedented abundance of funds. For experimental biology, for example, it was precisely the period between 1958 and 1963 which became the most exciting. The climax was the deciphering of the genetic code and the birth of molecular biology. However, despite the fact that the first report about the genetic code, by the American biochemist M. Nirenberg, had been presented at the Fifth Biochemical Congress, which took place in Moscow in August, 1961 (the first really large international scientific congress held in the USSR since

1935), Soviet biology was still dominated by Lysenko. A few months before the congress Khrushchev had reconsidered his decision made five years earlier and reappointed Lysenko president of the All-Union Lenin Academy of Agricultural Sciences. This represented a large step backward for the international reputation of Soviet science, which only a few months before had a new cause for pride—Yuri Gagarin's first manned space flight in April, 1961.

Gagarin's space flight was also a great achievement for the "chief designer," Korolev, who was responsible for the manned space program. We now know a great deal about Korolev, but in 1961 his name was unknown to the general public. The leaders of the nuclear and other defense-oriented research programs were "secret"—a curious practice inherited from Stalin's time. A few, like Igor Kurchatov, were excepted, however, when they became mainly administrators of governmental rank.

Khrushchev, who had unexpectedly included Kurchatov in the government delegation to Britain in 1956, in general continued Stalin's practice of suppressing the names of most prominent scientists in sensitive areas.

In his book *Khrushchev Remembers: The Last Testament* (41), Khrushchev tried to justify this practice by reference to the fear that foreign intelligence services could kidnap Soviet scientists. "In the case of international conferences," writes Khrushchev, "we often sent the second- and third-level experts, rather than the people in key positions. Thus, any kidnappers would be unable to get their hands on those few scientists who had a concrete, firsthand knowledge of our top-secret projects" (p. 59).

This anonymity of the key figures in space, nuclear, and certain other fields continued after Khrushchev's fall in 1964. The real name of the "chief designer" was acknowledged only when Korolev died in 1966—in a long obituary published in all central newspapers and signed by all leaders of the Party, government, and scientific establishments.

Korolev died at the age of sixty, when he was still full of energy

and ideas. He had for a long time suffered from a minor disorder—hemorrhoids. Finally an operation was recommended.
The operation was expected to be simple, and there was no
proper preliminary diagnosis. In view of the importance of the patient, the Minister of Health, B. V. Petrovsky, who formerly had
been an outstanding surgeon, but who had certainly had little
practice while a minister, decided to lead the team performing
the operation. While it was in progress, a small tumor was unexpectedly discovered. Neither the proper equipment nor supplies
of blood plasma had been provided for a long and complex operation. But Petrovsky decided to go ahead with additional anesthe-
sia. Damage to arteries, heavy loss of blood, and other complica-
tions which could have been easily overcome with the aid of such
modern equipment as the artificial lung-heart system made the
operation fatal, and Korolev died on the operating table. These
details were not published and the official bulletin referred to his
death "after a long and fatal illness." (In 1976, in articles devoted
to the seventieth anniversary of Korolev's birth, the accidental
nature of his death was acknowledged, but without details.)

The Beginning of Dissent in Soviet Science

The widespread repression of scientists and intellectuals during Stalin's time could create the impression that dissidence was
equally widespread among them. This was not the case. There
were several trends in this repression. One was political, the vic-
tim being of the wrong "class origin," formerly a member of a
non-Bolshevik party, and so on. Another resulted from the struggle between the younger generation, or pseudo-scientific groups,
and the older, more experienced scientists. But in most cases no
reasons were given and all the charges were fabricated.

Stalin was an anti-intellectual in many of his actions. He expressed his views clearly in the short speech he made in 1938 at
the reception in the Kremlin to celebrate the first expedition to
the North Pole, led by D. Papanin. In this he said that it was not

education or academic position that was important, but natural creativity, enthusiasm, and *novatorstvo* (unorthodox approach). In practice Stalin often supported badly educated or even primitive people, considering them to be "great" scientists because they had declared their intention of carrying out some extraordinary achievement. Sometimes this support was justified and the achievements of these unknown pioneers were later recognized throughout the world. Examples include Papanin's successful expedition to the North Pole, V. Chkalov's first transarctic flight from the Soviet Union to America, and A. Stakhanov's world record for coal output (using the special methods which started the "Stakhanovite" movement).

But very often Stalin's anti-intellectual support was given to semi-educated pseudo-scientists, like T. D. Lysenko and O. Lepeshinskaya. They then tried to establish a dominating position through ideological pressure and by using the apparatus of terror. It was just this cruelty and terror of the state machine which made organized dissidence in science impossible.

Individual cases of dissent among scientists were possible, however, and a few are known. One example was the refusal of academician Peter Kapitsa to participate in research related to the atomic bomb—an action which cost him his position and job, but not his life. Another was the strong opposition of academician V. N. Sukachev to Lysenko's forest planting methods in Russia's south and southeast. (This plan was an important element in Stalin's Program for the Transformation of Nature, adopted in 1948.) This kind of dissent was very risky for the individual, not only from the point of view of his scientific position but from that of his life and freedom as well.

Khrushchev's bold denunciation of the Stalin terror in 1956 and his rudimentary attempts to establish some legal justice in the country gradually stimulated the intellectual climate, including the scientific. Thoughts, ideas, scientific freedoms, and discussions all experienced a certain release. Representative groups of the scientific community began in some cases to oppose government and Party policy. But the inertia of fear aborted many such attempts.

Scientific opposition did, however, play an important role in changing the course of much that had already been approved by the Party. Cybernetics, for example, was rehabilitated as a useful science after Stalin's death, and many other new technological trends that had previously been suppressed struggled to the surface with the help of political pressure.

An important focus for scientific and political opposition arose in the agricultural and biological sciences over the Lysenko issue. Hundreds of scientists—not only biologists, but chemists, physicists, and others—united against Lysenko and against Khrushchev's support of Lysenko. This struggle was reflected in many official meetings and countless speeches by Khrushchev in which he attacked those scientists who opposed the "fruitful and revolutionary Michurin-Lysenko biology."

Khrushchev's decree in 1963 for an unrealistic 70–80 percent annual growth in the chemical fertilizer industry was challenged by a number of scientists. They wrote an open letter to the government and Party. The letter had some effect, and the "chemicalization" program was modified slightly.

Even Khrushchev's reasonable attempt to reorganize the Academy of Sciences in 1960 was strongly opposed. Members of the Academy and the Academy presidium refused to cooperate with the government. It took more than a year to settle the dispute. When it became clear that it was impossible to carry out the reform simply, the government decided to go ahead without the Academy's cooperation. The president of the Academy, A. N. Nesmeyanov, was forced to resign.

Awareness of environmental problems also started during Khrushchev's time. Scientific groups often opposed industrial projects on environmental grounds. In some cases the government made concessions and projects were altered. But more often, both during Khrushchev's time and after, the government was reluctant to alter its industrial plans for environmental reasons. The best-known case of this kind was the project for a big cellulose pulp mill on Lake Baikal in Siberia, which threatened the extremely pure water of this unique lake. The project, designed during Khrushchev's time, was carried out later despite strong scientific objections based on the damage that might be caused to the environment. Some changes were made, as a compromise, to prevent pollution of the lake, but it is impossible to know as yet how successful they have been.*

Khrushchev's initial measures in support of science—encouraging scientific exchange with the West, developing new science centers—together with his policy of "destalinization" and the rehabilitation of political prisoners won enthusiastic support among scientists.

During this period (1953–1957) he was even sensitive to demands about Lysenko. In 1956, Lysenko was dismissed from his position as president of the Lenin Academy of Agricultural Sciences of the USSR. He was reinstated in 1961 when he sided with Khrushchev in his quarrel with agricultural experts and scientists.

Khrushchev also abolished the annual Stalin Prizes, which had been awarded for the best achievements in science and technology. He introduced the Lenin Prizes in their stead. These were to be awarded once every two years, there would be fewer of them, and they would involve less money. This reform was, however, not very significant.

Khrushchev's real conflict with the scientific community

* Baikal is the largest freshwater lake in the world (when not only space but the amount of fresh water are compared). It is about 650 km. (400 miles) long and about 80 km. (50 miles) wide. What is unique is its depth—its maximal depth is about 2 km. (5,700 ft.), which makes Baikal also the deepest lake in the world. The scientists tried to protect Baikal not only because of its rich fish grounds and crystal-clean water. They argued that a shortage of fresh water in the world is inevitable and that Baikal is a strategically important enormous reserve of fresh water. This fresh water is a much more valuable asset for the USSR than the cellulose which is to be produced by the new industrial complex. Now it is clear that the construction of the plant there was a mistake, and anti-pollution designs to reduce the amount of organic material from the plant do not work efficiently because special pools where organic material is decomposed through bacterial processes are not active during seven months of the Siberian winter.

began to be felt after 1958–1959, when the "anti-Party" opposition had been eliminated and he had become the *de facto* dictator of the Soviet Union. In accordance with the long-standing tradition of the Communist movement, this meant that he now wanted to be thought of not only as a political leader—the First Secretary of the Party—but also as some kind of "super-scientist" (in the sense that Marx, Engels, Lenin, Stalin, and Mao Tse-tung were all officially considered "great scientists" and the ultimate authority). The idea that the leader of the USSR and of the Communist movement as a whole had the best mind was still prevalent in 1958. The temptation to acquire such a reputation was too much for a person in a position of almost absolute power.

Khrushchev's conflict with the scientific community contributed to his downfall in 1964. Some episodes in this conflict—such as his closing of the Moscow Timiriazev Agricultural Academy, his support for Lysenko's pseudo-science, and his attempt to reorganize the Soviet Academy of Sciences into a Committee of Science in 1964—were mentioned by Suslov in the report he made to the Party plenum in October, 1964, as among the reasons for Khrushchev's dismissal.

Two tragic episodes which exposed the explosive tensions between Khrushchev and the scientific community were of particular importance. They strained Khrushchev's relations with two groups of very influential scientists. Both these groups—the nuclear physicists and the spacecraft and rocketry technologists—were the elite of the elite. They were also essential for the country's strength, probably more essential than Khrushchev himself.

But it was the suppressed geneticists who originally started the conflict between Khrushchev and the nuclear physicists. In 1955–1956 they made a number of attempts to arouse the physicists to the dangers of radioactivity for humanity as a whole. This underground propaganda, which emphasized the genetic damage caused by radiation and the importance of classical genetics to control it, was rather successful. By 1956 several hundred signatures had been collected on an appeal calling for the restoration of genetics and of radiation genetics in particular. The tsar of the nuclear physicists, Igor Kurchatov, handed this appeal in person to Khrushchev. Khrushchev was furious, but he could not touch Kurchatov, who had too strong a backing. Finally, Khrushchev made some small concessions. One was the removal of Lysenko from the presidency of the Agricultural Academy.

But in 1958 Lysenko returned to favor and got his revenge. His "victories" included the dismissal of the whole editorial board of the anti-Lysenko *Journal of Botany,* the dismissal of Nikolai Dubinin from his position as director of the rival Institute of Cytology and Genetics in the new Novosibirsk center of the Academy, and the dismissal of the anti-Lysenko chairman of the Soviet Academy's Biology Division, academician V. A. Engelgardt.

A tragic catastrophe occurred at the end of 1957, which made nuclear physicists extremely sensitive to the radiobiological and genetics issue. The government finally introduced measures legalizing classical genetics, at least in radiobiology, radiology, and medicine. The complex new rules and regulations governing the study of genetics and of the danger of radiation were radically changed. Lysenko's only remaining power base was in agriculture.

This catastrophe,* which altered the mood of the nuclear physicists, could have been foreseen, although a number of nuclear experts did not believe it was possible. For many years nuclear reactor waste from several reactors had been buried in special trenches in a deserted area of the Cheliabinsk region in the South Urals. The waste was not buried very deep and was not properly diluted. Nuclear scientists had often warned that this primitive method of waste disposal was dangerous, but nobody took their views seriously.

At the beginning of the Soviet atomic program, priority was

* I have already described this disaster briefly in 1976 in my article in *New Scientist* (42). This article caused a lot of discussion, because the tragedy, which had happened about twenty years before, was quite unexpectedly not known about in the West. For that reason I am including here some additional details. See also Appendix II.

given to the rapid buildup of an arsenal of atomic weapons. Many problems, including that of nuclear waste, were faced only as and when they emerged. The solutions were not always the best. The deadlines for atomic weapons development had been so tight that the first reactor, tested in 1946 in Kurchatov's own laboratory, did not have any protective radiation shield. The tight schedule meant there was no time to build proper protection; the operation team simply moved underground as soon as the radiation areas became about half a mile long (29). The concrete protective shield was built later, when the reactor passed into the experimental stage.

The first military reactors, which started to produce plutonium in 1948, were built to better plans. The problem of waste released after processing was not solved, however, before the waste became available. Details about the disposal of waste are not yet clear, but for a long time it was buried in primitive underground shelters, probably highly concentrated and not very deep. It is clear now from some research (45) that the remaining plutonium could have been absorbed selectively by the soil, with the result that its concentration could have grown to a dangerous level. The accidental soaking of this soil—for example, by heavy rain—could trigger a nuclear chain reaction. The energy released by such a process might throw up some kind of "mud volcano." (This kind of danger was discovered in the United States at the Hanford nuclear waste reservation, and the American Atomic Energy Commission started a project in 1972 to prevent any such occurrence.)

The kind of process that caused the explosion of a waste disposal site in the Cheliabinsk region was never reported. This explosion, however, carried the heavily contaminated soil over more than a thousand square miles in the lake district lying between the two big industrial cities, Cheliabinsk and Sverdlovsk. Strong winds blew the radioactive clouds for dozens of miles. It was difficult to judge the extent of the tragedy immediately, and no evacuation plan was put into operation right away. Many villages and towns were ordered to evacuate only when the symptoms of radiation sickness were already quite apparent. Tens of thousands of people were affected, though the real figure has never been made public. Probably many hundreds died quickly, thousands more slowly, and the full impact of the tragedy will probably never be known. The whole area where the accident occurred is still considered dangerous today and is closed to the public. A number of biological stations have been built on the edge of the affected area so as to study the radioactive damage to the plants and animals. This is certainly the biggest radioactive field in the world.

The irradiated population was sent to many hospitals. But no one really knew how to treat the different stages of radiation sickness, how to measure the radiation dose received by the patient, or how to predict what the effects would be, both for the patients and for their offspring. Radiation genetics and radiology could have provided the answer, but neither of them was available. There was no laboratory in the whole country which could make a routine investigation of chromosome aberrations (the most evident result of radiation exposure); bone-marrow stocks did not exist; there was no chemical protection against radiation exposure available for immediate distribution. Many towns and villages where the radioactive level was moderate or high, but not lethal, were not evacuated, or evacuated only later. The medical observation teams established in them were not properly equipped to carry out tests.

The nuclear scientists were tremendously shocked by all this, and their opposition to Khrushchev's anti-genetic stand became too strong to resist. If the Russian government does not often understand the language of reason, it does understand the language of a catastrophic emergency.

The nuclear physicists were now well aware of the real dangers of radiation and of the real consequences of atomic explosions. When, in 1961, Khrushchev announced the testing of two 100-megaton nuclear bombs in the atmosphere for purely political reasons, he met with strong resistance from the nuclear scientists. Soviet physicists were no longer just an obedient group of experts. Their strong opposition to government policy contributed considerably to the final agreement to end all nuclear explosions aboveground.

It is quite possible that the Soviet government's short-lived decision to suspend all tests of nuclear bombs, declared at the very beginning of 1958, was in some way related to the nuclear disaster in the Urals area. It was probably necessary to shut down plutonium-producing plants temporarily, and this was a good excuse for Khrushchev's unexpected move. The nuclear physicists engaged in the nuclear arms program were satisfied by this decision. The military experts, though, tried to apply pressure in the opposite direction; they had some new bomb prototypes to explore.

When, in November of 1958, Khrushchev decided to resume testing on the ground and in the air, the research elite of nuclear science were very unhappy. Andrei Sakharov, who was most prominent in Soviet H-bomb projects, was the first to protest strongly against the new test program.

The difference in attitude of scientists and pragmatic politicians with regard to the value they placed on human lives is well illustrated by the dialogue between Khrushchev and Sakharov, which Khrushchev reconstructed in his book (41).

Literally a day or two before the resumption of our testing program, I got a telephone call from Academician Sakharov. He addressed me in my capacity as the Chairman of the Council of Ministers, and he said he had a petition to present. The petition called on our government to cancel the scheduled nuclear explosion and not to engage in any further testing, at least not of the hydrogen bomb: "As a scientist and as the designer of the hydrogen bomb, I know what harm these explosions can bring down on the head of mankind."

Sakharov went on in that vein, pleading with me not to

allow our military to conduct any further tests. He was obviously guided by moral and humanistic considerations. I
knew him and was profoundly impressed by him. Everyone
was. He was, as they say, a crystal of morality among our scientists. I'm sure he had none but the best of motives. He was
devoted to the idea that science should bring peace and pros-
perity to the world, that it should help preserve and improve
the conditions for human life. He hated the thought that
science might be used to destroy life, to contaminate the atmosphere, to kill people slowly by radioactive poisoning.
However, he went too far in thinking that he had the right to
decide whether the bomb he had developed could ever be
used in the future [p. 69].

Sakharov also reconstructed this dispute in his book (31), and he
dates the beginning of his political dissidence from this episode.

The large area in the South Urals contaminated by radioactive
isotopes (mostly strontium 90 and cesium 137, which have half-
lives of about thirty years and need several hundreds of years to
be completely destroyed) continued to grow after 1957, because
of so-called "secondary distribution," through soil and wind erosion and biological processes. All ecological and radiobiological
research carried out in this area since 1958 was secret. However,
some works have been declassified since 1966–1967,* on condition that the effects resulting from radioactive contamination of
the soil, plants, animals, lake population, and so on are published without indicating the exact location of the experimental
area or its total size.

The other line of dissent and resistance against the political
use (or, better, "overuse") of science developed in space research
after 1960. Khrushchev was especially proud of space research.
It was closely related to military rocket technology, and many
space programs in the USSR, as well as in the United States,
were related to political prestige. This was probably natural, but

---

*A review of these works is outside the province of this book, but some refer-
ences are given in Appendix II.

only up to a point. Khrushchev overstepped this mark too obviously and spoiled his whole relationship with the extremely influential space and military groups. They started to resist political
pressure and wanted a significant role in the decision-making
process. They got it, but only after another irreparable catastrophe.

Relations with the United States became rather tense in 1960
after the failure of the summit meeting, due primarily to the in-
cident concerning the American U-2 spy plane. In October of
that year Khrushchev decided to head the Soviet delegation to
the General Assembly meeting of the United Nations. He and
some other heads of the East European countries made the trip
on the ship *Baltika*. Always obsessed with the idea of showing
the Americans Soviet superiority in at least some areas of technology, Khrushchev issued a directive that a Soviet rocket to the
moon should be launched to coincide with the time of his arrival
in New York. It would be some kind of space *salyut* for the arrival
of such an important Communist group to the United States.

Marshal Nedelin was made responsible for the project. The
elite of Soviet rocket technology were, of course, at the "cos-
modrome." However, when the start was ordered and the button
was pressed, the ignition did not work. Something was wrong.
According to the safety regulations any inspection could only
take place after the fuel had been removed from all the stages of
the rocket. This was a long process and would mean postponing
the whole spectacle. Marshal Nedelin, who was under an obligation to fulfill the ambitious order, took an irresponsible decision—
to investigate the fault immediately and make the necessary cor-
rections. The special ladders and platforms were moved to the
rocket, and dozens of engineers and experts began to explore the
different parts of the multistaged rocket system. Nedelin himself
coordinated the efforts. Suddenly the ignition of the second-stage
rocket started. The rocket fell and exploded. All the men and
women in the area were killed. They were some of the best representatives of Soviet space technology. This tragedy was not the

only event to make the space technologists aware of the dangers of political *salyuts*. The government's attempt to hide the real story meant that the tragic death of many prominent scientists and technical experts passed without even short obituaries.

Khrushchev describes this catastrophe in his memoirs, *Khrushchev Remembers* (41), but in quite the wrong light. Praising the chief rocket designer, academician M. K. Yangel,* for his contribution to military and peaceful space research, Khrushchev wrote as follows:

> Chief Designer Yangel just barely escaped death in a catastrophic accident which occurred during the test of one of our rockets. As the incident later was reported to me, the fuel somehow ignited, and the engine prematurely fired. The rocket roared up and fell, throwing acid and flames all over the place. Just before the accident happened, Yangel happened to step into a specially insulated smoking room to have a cigarette, and thus he miraculously survived.
>
> Dozens of soldiers, specialists, and technical personnel were less lucky. Marshal Nedelin, the Commander in Chief of our missile forces, was sitting nearby watching the test when the missile malfunctioned, and was killed [p. 51].

*M. K. Yangel was a German expert, colleague of Werner von Braun, the main designer of military rockets in Germany. Von Braun, after the end of the war, started to work in the United States with most of his colleagues. The Soviet army was, however, able to capture many rocket engineers in Germany as well. They were put to work in some prison research centers. Many of them later returned home when the German Democratic Republic was established. Yangel, however, decided to stay in the USSR. He became a close collaborator of S. P. Korolev, and received many awards and promotions. He was also elected a member of the Academy of Sciences of the USSR.

After Korolev's death in 1966, Yangel became the "chief designer"; however, he died in 1971. Yangel's name is not so well known as the name of Korolev, who received great publicity after his death. The names of the chief military experts, as I have already mentioned, are usually kept secret, and they receive all their honors and awards secretly. However, when they die their names and biographies are made public. This was easy to do for Korolev, who had a remarkable biography. For Yangel, who had worked in Nazi Germany before 1945, posthumous publicity was not possible. Up to now his name and his contribution to the development of Soviet space technology have not been widely known.

In the official press, however, the death of M. I. Nedelin was reported as being due to a "plane crash" (October, 1960), while nothing was said about the death of many experts, some of whom had been high-ranking scientists and technologists. The whole story was in fact quite different from the way Khrushchev described it, and he himself was mainly responsible for the accident. The duplicate rocket was later launched and hailed as a great achievement. But this could not heal the wounds of those who lost their relatives, friends, and colleagues.

These are just a few examples that show how Khrushchev began to lose the confidence of Soviet scientists. They also show that the dissent was not only among the rank and file, but also at the level of the highest scientific elite. In his memoirs Khrushchev acknowledged (with numerous errors) that he had been wrong in his attitude to many important scientists and creative members of the intelligentsia. He described his conflicts with academician Peter Kapitsa and Andrei Sakharov, and with some prominent figures of the artistic elite. In some cases the quarrels were over minor issues; for example, when he would not allow Kapitsa to travel abroad. The culmination of this conflict was a special Party plenum held in June, 1963, to discuss ideology and ideological orientation in science, literature, and art. This plenum was reminiscent of the notorious decisions taken by Stalin and A. Zhdanov on the superiority of ideology and its relevance to all aspects of science, literature, and art. Their decisions had initiated the repressive measures against intellectuals in 1946.

However, the conflict between Khrushchev and the top-ranking scientific elite not only added fuel to the anti-Khrushchev move made by the Party Presidium in October, 1964; it also created a unique situation in which the highest scientific support became possible for *political* dissidents, who had never belonged to the elite.

Political dissidents within the scientific community were usually at the lower levels of the scientific hierarchy. The young and junior scientists, not the privileged academicians, tried to explore some political alternatives and ideas during the postwar period. Their fate under Stalin was usually tragic. During Khrushchev's time, the first wave of arrests among young scientists and student dissident groups came after the military intervention in Hungary in October, 1956. These arrests are not well known, because the rehabilitation of millions of victims of the Stalin terror was under way at the same time. When millions were being released, the hundreds of new arrests could easily pass unnoticed.

These young dissidents had been brought into existence by the new policy of "destalinization." However, they wanted more serious reforms in society, a more stable democratization. They were stunned by the details of the Stalin terror; they wanted a more complete investigation and the punishment of those others who had also been guilty of such crimes. In 1956–1957 there were few such dissidents, and they were isolated from almost all groups in society—from workers and peasants because of the lack of any means of communication; from higher sections of the intelligentsia because of the latter's privileged, elite position and their satisfaction with the half measures of Khrushchev's regime toward liberalization. Too many of the intellectual elite of 1956–1960 were still famous from Stalin's time; they had been in some way responsible for the "Stalin cult" and had been directly or indirectly involved in different purges. The situation changed around 1962–1963. A certain cooperation grew up between the democratic political opposition and the scientific elite opposition.

The prestige of the numerous state orders, the signs of political recognition, the titles, prizes, degrees, and even high positions were all tremendously eroded and devalued. By 1963 it was considered ridiculous to display such orders or medals on one's suit. These titles, orders, and prizes had been mostly won during Stalin's time, and the question naturally arose: What had one done to receive the title Hero of Socialist Labor? Was it for the development of a good new variety of wheat, of a new version of the nuclear bomb, or had one overfulfilled the construction program for a hydroelectric dam being built by slave labor from the prison camps? The same questions could be asked about prizes and degrees. As a rule, the most decorated scientists came from the Lysenko camp, with himself well in the lead—seven Orders of Lenin, Hero of Socialist Labor, several Stalin Prizes, full membership in three academies (USSR, Ukraine, and Agricultural), member of the Academy presidium, director of the Institute of Genetics and of the Gorky Leninskie Experimental Station, deputy of the Supreme Soviet, and so on.

The devaluation of the scientific hierarchical pyramid made closer contacts between the younger politically active groups and the more honest representatives of the elite much easier. In many cases the high-ranking elite scientists were themselves looking for such contact. Right up to 1957–1958 the politically orientated young dissidents would have considered cooperation, free meetings, or any kind of friendly relations with such scientific celebrities as academicians T. Tamm, P. L. Kapitsa, A. D. Sakharov, N. N. Semenov, V. A. Engelgardt, I. L. Knunianz, and A. I. Berg quite unthinkable. But by 1962–1964 links had started to appear. Not only did cooperation and friendship between the two generations become possible, but the older and more privileged one gave direct support to the political dissent of their younger colleagues. They gave financial aid to help organize the *samizdat* network, they sometimes made facilities available for safeguarding and reproducing *samizdat* works, and they also strongly encouraged and defended those in trouble. In a few cases members of the highest scientific elite became outspoken political dissidents themselves—the case of Andrei Sakharov is one of the best-known examples. But this happened later, not during Khrushchev's period.

Chapter 7

The Main Aspects of the Development of
Science and Technology During the First
Stage of the Post-Khrushchev Period
(1965–1971)

The political tides in the history of the USSR are often related to the Communist Party congresses. Therefore, in this chapter I will consider those events leading up to 1971. In 1971 the Twenty-fourth Party Congress decided on more radical changes to establish *détente* in international affairs. This new *détente* policy had a serious influence on the position of science and its effects on scientists. In 1971 the "duplicating" trend in technology and science slowly started to disappear and to be replaced by a "cooperative" trend. This meant dramatic challanges for science, which had to be diverted from "parallel competition" with (perhaps, better, "repetition" of) Western science toward a gradual, cautious, and controlled *integration* with international science and world technological progress. This new trend was not, however, spontaneous. The way had been prepared by the events of the 1965–1971 period and mainly by the obvious fact that Russia had not won the over-all technical and scientific competition with the

West. To be frank, the USSR was behind in almost all the sophisticated areas of modern technology and science. The "technological gap" discussed in the previous chapters of this book had not been reduced. This had been clearly demonstrated by the spectacular success of the United States' first expedition to the moon. The television pictures of Americans walking on the moon's surface and safely returning home made a tremendous impression on official Soviet science policy.

Immediate Changes after the Fall of Khrushchev

On October 12–13, 1964, during the meeting of the Central Committee of the Communist Party which decided to dismiss Khrushchev, some of Khrushchev's errors and wrong action concerning science and scientists had been mentioned by his opponents. Chief among them was the Lysenko issue. Khrushchev ignored all the appeals from prominent scientists about Lysenko, but other members of the Presidium of the Party seemed to be well informed about the pseudo-scientific character of Lysenko's ideas and his failures in practical agriculture. Lysenko was not dismissed from his many key positions immediately, but the very fact that he was strongly criticized at the Party Central Committee meeting was enough to start an open scientific discussion in the national press. It was open criticism that very soon decided Lysenko's fate. The Academy of Sciences of the USSR in 1965 made a special investigation of the different aspects of Lysenko's work, and he was later dismissed from his most prominent positions. However, he retained the title of academician and the position of Chairman for Science in the Academy's Agricultural Experimental Station not far from Moscow was also reserved for him. He kept this post until the end of his life in 1976. Considering his almost dictatorial powers in Soviet biology for many years, it was a minor position for Lysenko, but it was by no means insignificant according to normal standards.

The Experimental Station of the Academy of Sciences of the USSR is a big research center with more than 500 employees. The whole settlement represents a small satellite scientific town near the last Lenin residence, Gorky Leninskie. Here Lysenko was a king, with his research team, several laboratories, cattle farm, and many experimental fields. However, no serious research work was published from this "scientific center" during his last ten years.

Intensive efforts had been made at the end of 1964 and 1965 to restore genuine genetic research in the USSR. Many laboratories, departments, and chairs at universities and agricultural higher colleges were opened. Substantial financial help was given for radiobiology, radiation genetics, medical genetics, and many other fields of biology.

The logic of this very positive development in Soviet science made it clear that the artificial division of scientific disciplines into "socialist" and "bourgeois" was wrong and had harmed Soviet science by alienating it from the mainstream of world science. The effect of the persistent and long-time application of the wrong principle had been felt not only in biology but in many other areas of science. In biology, the problem, however, was not the *gap* in its development, but the real vacuum of knowledge in many fields of research created by thirty years of Lysenko domination. In other sciences the problem had not been so acute. The continuing progress in space research, where the new generation of spacecraft was available for more complex experiments; the impressive development of nuclear physics, the new accelerators of elementary particles, "plasma" nuclear research; and the new generation of civil and military aircraft—all these specially supported fields of Soviet science and technology helped to create a positive image of Soviet science at least for internal use and official declarations. In other areas this image was not easy to maintain.

The new Party and state leadership did not initially consider science its priority problem. The country, after Khrushchev's hectic final years in power, was left on the verge of economic bankruptcy, and the first priority was industry and agriculture.

There was much reverse reorganization during 1965–1966: the restoration of centralized ministries, industrial reform, the restoration of unity in administration within the Party and government networks. There was also the enormous task of raising the agricultural output of essential products. Khrushchev's fall naturally gave more influence to the conservative "Stalinist" elements in the leadership and opened some rudimentary possibilities for political struggle. Democratic ways of creating opposition trends from the "right" or from the "left" did not exist in the Soviet hierarchy,* but the ruling group was pressed from both directions. The old Party bureaucracy, military and security systems, and the provincial Party bosses pressed strongly from the right. Top scientific advisers and the scientific elite with its technocratic representatives in the Party and government systems pressed from the left, but this pressure was less influential. The new post-Khrushchev leadership was not yet stable, and unofficial discussion about the provisional character of the Brezhnev-Kosygin-Podgorny group was quite common in 1965–1967. In such a situation of muted political struggle for power, the development of science became more autonomous, and the top administrators—presidents of academies, chairmen of state committees on science and technology, heads of different complex programs such as space and nuclear energy—became more powerful and independent of political leaders. The scientific elite became more politically conscious, and the collaboration between the highest-ranking scientists and the politically charged and politically better-educated younger groups of dissident scientists closely connected with the "democratic movement" became stronger. Top scientists looked for political connections with

*"Left" and "right" in Soviet unofficial political language are used in a very different sense from that in the West. The "right" are Communist hard-liners; the "left" are more liberal and democratic.

prominent writers, artists, film directors, actors, and other intellectuals, and this union reacted strongly against attempts of conservative groups to rehabilitate Stalin and reintroduce their ideological dominance. It was not accidental that the strongly worded letter to the Party Central Committee protesting the attempts to politically rehabilitate Stalin was signed in 1966 by twenty-five highly placed figures in science, literature, and art. All who signed this letter were world famous and at the same time loyal to the Soviet system. Among them were academicians P. L. Kapitsa, L. A. Arzimovich, M. A. Leontovich, A. D. Sakharov, and I. E. Tamm. It was also signed by the famous writers K. Paustovsky and K. I. Chukovsky and by other artists and actors, including Maya Plisetskaya, the world-famous ballerina, and M. Romm, the film director, noted for his films about Lenin. This appeal was important and influential. It was made public in *samizdat* before the Twenty-third Party Congress, and it soon became clear that others were ready to sign this letter as well; many new academicians and prominent scientists made open statements that they attached to the message. The appeal got certain attention, and Stalin's name, which had been mentioned too often in the official press in 1965 and in early 1966, was not mentioned at all in the official report of the Party Congress and in the speeches during discussions held there. It was also not accidental that when A. Solzhenitsyn, already under open harassment, tried to find protection and an audience for oral readings from his forbidden works, he began to receive invitations for meetings from important research institutes. The first such meeting was organized at the Kurchatov Institute of Atomic Energy. Although Kurchatov died in 1961, this institute, a type of headquarters for all nuclear research in the USSR during Kurchatov's directorship, remained as the main scientific center.

Scientists generally had been active in protesting against the many political trials during 1966–1968, and many prominent names could be found as signatories in appeals for the release of Sinyavsky and Daniel. I talked with academician P. L. Kapitsa in 1967, and the discussion quickly turned to political problems. Kapitsa made an interesting observation: "To be able to maintain democracy and legality," he noted, "it is absolutely necessary for a country to have an independent institution to serve as an arbiter in constitutional problems. In the United States this role is reserved for the Supreme Court and in Britain for the House of Lords. It seems that in the USSR this function falls morally on the Academy of Sciences of the USSR."

Khrushchev's "duplication" policy inevitably concentrated the attention of the research community on practical goals. It was new technology and new methods which had to be copied and used in the USSR. During this period the industrial research network received priority, and the Academy of Sciences was forced to give up its applied research institutes (constituting more than half of the Academy structure), which had been dispersed all over the country and located in places where appropriate industrial production or developments had been concentrated.

The respectability of more general research work was slowly reestablished after 1965, especially after the end of pseudo-sciences of the like of Lysenkoism. The Academy's growth after 1965 was mainly associated with the applied branches again. Practically useful works and the "state-need" orientation were the best methods to promote good relations with the new Party leadership and government, and were also helpful for receiving all possible administrative financial and moral support.

The State Committee for Coordination of Scientific Research Work, established in 1961 and transformed later into the more powerful USSR Council of Ministers' State Committee on Science and Technology, was initially established to coordinate and to lead many different independent industrial and academic research networks. However, it did not become the highest administrative authority in the Soviet hierarchy of science and technology, as often could be seen in the different structural schemes or organization of Soviet science (44–46). This committee, a purely bureaucratic superstructure, was not able to employ prominent scientists possessing the necessary knowledge and authority to head its divisions and departments and to coordinate the research programs in many fields. In order to survive (all bureaucratic systems want to survive as soon as they are created), this committee elected to supervise and coordinate some of the more prestigious (and expensive) projects, like large complex programs of military importance (rocketry, nuclear science, aircraft, shipbuilding), some problems in technology, in medicine (cancer), and in general sciences (radiobiology, molecular biology, etc.). The scientific projects had been divided into "state important" and "others." "State important" became better financed and better connected with the different industrial branches through the State Committee. However, this was a functional, not a leading, role. The State Committee became just one more link between science and industry. If it were to be abolished now or in the future, there would be hardly any change, as all research units in the USSR are under government control by the ministerial, regional, or academic systems.

The End of "Duplication": New Objectives for Scientists

Khrushchev's ideas about the possibility of rapidly copying Western technology in order to make the "great leap forward" proved wrong in many areas of science and technology. As I have already tried to show in my book on scientific cooperation (9), the idea of a contradiction between capitalism and scientific progress, one of the dogmas of the Leninist version of Marxism, did not fit the situation of postwar scientific and technical development.

Science and technology in the West developed so quickly that the process received the name "Scientific-Technical Revolution." In such a situation the practice of slow copying created the conditions for a permanent and increasing lag. World technical progress had taken so many forms in different countries that the problem of proper selection and testing in order to find the "best" was difficult and slow. Up to now the process of copying had turned into the copying of the obsolescent and obsolete. In the USSR the average time needed to introduce a new technical model into production was longer than, for example, in the United States or Japan, and this situation contributed to the technological gap created during previous years. The process of analyzing the models of equipment from different countries and assimilating the best from them proved ineffective because the hybridization of sophisticated equipment was not always possible. I have already mentioned that the Institute of Medical Radiology in Obninsk had a special workshop and engineering bureau to develop modern radiological and roentgenological equipment. Samples were brought here from Japan, Italy, Britain, France, Germany, and the United States; a worldwide collection of different designs. They were different in that each sample was unique yet perfect. The team of institute engineers was not able to design anything really synthetic and modern from the study of all this apparatus. Similar developments occurred in the case of electron microscopes. Although the Soviet models were obsolete, the special team which tried to design a new model finally gave up the attempt to find a successful hybrid between the Japanese Hitachi and the German Siemens microscope. Their decision after 1966 seemed simpler—the Japanese microscope was more economical (but perfect), and it was selected as suitable for the USSR.

In 1966 the USSR joined the International Convention on Patents, Licenses, and Inventions. Therefore, the whole Japanese plant for the production of electron microscopes was purchased. But even this approach was not enough. Soon it was found that not only the equipment but the expertise of Japanese engineers was necessary for the manufacture of a good product. When the plant was finally in operation, it nevertheless did not produce microscopes of the same precision and fidelity as the original Japanese models, although the design was the same. Japanese white-

and blue-collar workers were needed to achieve the perfection of the genuine Japanese model. A Japanese firm offered to send assembly-line workers and technicians to various parts of the enterprise, who would work side by side with the Russians and teach them. But the Soviet officials who had so easily decided to purchase the Japanese plant, and had accepted the technical assistance of Japanese engineers and experts, were reluctant to employ Japanese blue-collar workers. When I learned about this problem, the offer from Japan was still under consideration (1972), although the electron microscopes produced in the USSR were of an inferior quality. Several research institutes using electron microscopes insisted on purchasing them in Japan, and would not use the "Japanese model" made by Soviet workers. I assume this problem has been settled by now.

The purchase of the Japanese plant, as well as the purchase of an Italian Fiat plant to produce cars, are just two examples of the slowly dying duplicating trend. After the Soviet Union joined the International Convention, in 1966, free duplicating of new technical or industrial ideas was more difficult. In science, however, this trend did not completely die away. Duplicating research in science is not controlled by any agreement, and parallel research sometimes occurs within a single country. The structure of a certain protein, for example, can be established and sequenced by three separate groups independently. There is nothing wrong with this, except possibly the waste of money. Different groups can pursue their work by using different methods and may find a few details others have missed, and the final discovery will be more valid when they have compared their data. Such an approach to duplication would have been welcomed by Russian scientists as well, as long as they were permitted absolutely free communication with their foreign colleagues, but because free cooperation had not yet been granted, parallel work by Russian scientists was, in most cases, delayed, and the repetitive duplicate study had little value when it was later published. This loss of time and resources in the continuation of copying trends was not as clearly evident in biology or chemistry as it was in technology or experimental physics. As soon as the Americans started to lead in two prestigious and expensive fields of science and technology—space research and research on nuclear physics with the use of particle accelerators—Soviet science was forced to recognize the advantages of scientific cooperation. The successful moon expeditions (the Apollo program) moved the United States forward in space technology and science. It was possible for the Russians to repeat this performance, but the price of such repetition was too high and the political and scientific advantages too small. The same was true of the competition with accelerators. The Russians had led the field before 1968, when the most powerful accelerator (which cost about one billion rubles) was completed at the science town of Pushchino. Only a few papers on this great machine were published before the Americans built a bigger accelerator in California, which cost them probably two billion dollars. The benefit to physics of both pieces of equipment was about equal, but higher power and speed of acceleration gave the Americans a clear advantage. In order to take a new lead, Russian physicists would have to spend five or six billions, more than half of their whole scientific budget. This was impossible. The race in space and on the earth's surface was over. (Military technology has a very high price, too, but this is inevitable, as nobody wants to share military secrets.)

When the continued race in the peaceful sciences and technologies became too expensive, the USSR simply could not keep up. In 1971 this was all too clear. For the sake of the country's economic progress it was necessary to start real cooperation with the Western countries. The new turn in Soviet science and technical development was inevitable. There were many domestic reasons for this. A new policy had to be declared sooner or later, but neither the Russians nor the Americans were able to find an ideologically acceptable word in their languages. They finally compromised and began to use a French word. This has now become the most famous, most ambiguous, most often used word in the world of international politics—the word *détente*.

Before we study the impact of the new political climate on Russian science, we should review the situation in Soviet science up to 1971. This gives us the possibility of analyzing the changes made in more recent times, changes directly associated with the *détente* policy, although there are still quite a few both in Russia and in the West who continue to insist that nothing has really changed within the Soviet Union since the implementation of the *détente* policy. Such a conclusion can be reached either by optimists who want too much too quickly or by pessimists who think that everything in the USSR has always been bad and nothing can be improved. Alongside them are many pseudo-experts and dishonest observers who easily ignore facts that conflict with their own theories. Arguments are not enough in such polemic; it is necessary to consider facts.

The Conflict between New Objectives for Soviet Science and Previous Forms and Conditions of Research Work

The gradual end of the "copying" approach and the "copying" psychology was not achieved by one or two top official decisions or by the swift reorganization of scientific and technological establishments. The analysis of foreign advances, the worldwide assimilation of scientific and technical literature on one subject or another are unavoidable stages in any research. If the simple "copying" trend in research depends upon administrative measures—the facility for rapid reconstruction, small modification and application of slightly changed or "synthetic" models, methods, or processes in Soviet industry or agriculture—then the new approach demanded different methods. In a situation where a wide qualitative gap in many scientific and technological areas still existed, the copying trends did not disappear immediately. The new possibility of buying a license for the production of a certain piece of equipment decided upon as superior very often meant that the methods of production of such a model had to be developed independently. It was naturally impossible to buy from abroad the means of production of every new model. There were many failures in such commodities, even though the licenses had been bought and the technology of production known. I will try to explain the newly developing situation in connection with scientific work. (Similar situations certainly existed in many industrial areas as well.)

From the beginning of the 1960s, serious biochemical research had become impossible without new efficient methods of fractionation of macromolecules based on special materials (ion-exchange chromatography, absorption chromatography, ultrafiltration, affinity chromatography, gel filtration, polyacrylamide electrophoresis, and many others). I do not want to complicate this description by using specialized terminology. These methods became absolutely essential for the purification of enzymes, peptides, hormones, and different proteins; they paved the way not only for new advances in theoretical biochemistry, but also for substantial developments in many practical fields such as pharmacology and the production of enzymes, food, polymers, and so on. They were used for all analytical methods, from medical-diagnostic biochemistry to analyses of the moon's dust.

Soviet science and industry needed the products of such fractionation in large quantities, and the methods of their preparation were available. However, when the organic chemical industry began to produce "Made in the USSR" ion-exchange materials or Soviet-made compounds for polyacrylamide gels or gel filtration, they were never (at least until 1973) of the same quality as the Swedish, German, or American products, and it was difficult to obtain the same results with the use of the Soviet versions of similar products. This kind of quality difference could be found in the Soviet version of "pure" enzymes, proteins, special chemicals widely used in medical diagnoses such as phytohemoglutenin, and others. In industry, where all these substances were used in large amounts, the Soviet-made products were, nevertheless, widely used (for cleaning of industrial waters, pharmacological production of enzymes and hormones, and many other processes). The country could not afford the imported products when tens of thousands of tons of synthetic materials were necessary.

The quality of the results at the beginning were not as high, but it was a start. In scientific research, however, foreign materials were absolutely necessary, and it finally became possible to import them for research purposes. Soviet industry and science, with their new freedom to introduce modern foreign technology legally (unlike the technological piracy of Stalin and Khrushchev's time), quickly eliminated the pattern gap of consumer and industrial commodities. However, the *quality* gap was more difficult to remedy, and the problem of quality remains unsolved in many areas.

It is interesting to ask why Japan, for example, which had also built up its postwar industry and economy on the principle of assimilation and had made use of the scientific and technical achievements of many other countries, could produce the same equipment not only more cheaply but *better* than in the country of its origin, while the situation in the Soviet Union is so clearly the opposite. At least one explanation is obvious: Japan produces mostly for the export market, and the Soviet Union produces for the internal market, where the pressure of competition is nonexistent. This difference is an important one, but it is not the only reason for the low quality of many Soviet-made products. As Soviet science was gradually relieved of its obligation to imitate the West, research centers and units became oriented toward the development of *new* solutions, *new* discoveries, and *new* methods. This new orientation produced few results during the first five to six years, in spite of the growth of the scientific establishment and a larger research budget. The poor performance in many areas was mainly the result of the same disease—scientific *isolation*. In 1965–1971 this isolation was not as prevalent as it had been in Stalin's time, when access to foreign journals was difficult and correspondence or direct contacts with foreigners was forbidden. There was some improvement under Khrushchev, and in 1955–1964 foreign exchange became restrictively available and the flow of scientific information from abroad increased. However, *free* exchange meant freedom to copy, and the division of science into "Soviet" and "bourgeois" schools continued. Exchange improved still further in 1965–1971, but not enough. Perhaps if the new "freedom" had begun in 1937 or 1949 instead of in 1970, Soviet science and technology would have advanced more rapidly. But in 1970 world technology and science had already achieved so high a level of sophistication in so many areas that it was impossible to eliminate the gap by reading journals or exchanging papers. The "new freedom" should have been more than the free exchange of knowledge; it should have been an exchange of experience, with cooperation and integration in research as well. But the Soviet system was not yet ready for this level of cooperation.

RESTRICTIONS ON FOREIGN TRAVEL FOR SOVIET SCIENTISTS

For the Soviet citizen foreign travel is a privilege, not a right, and it is hedged about with a complicated system of arrangements under political control at all levels.* These restrictions have serious implications not only for science but for the political and cultural development of the USSR at large. It is especially difficult for an individual to obtain permission for a professional trip or a scientific meeting abroad.

The published figures on Russian citizens visiting foreign countries every year are grossly inaccurate. They include repeated visits by the most important scientists and make no distinctions between visits to Western nations and to those of the

---

* The difficulties of foreign travel for Soviet scientists and other citizens have been described in a previous work (9). The situation which I described was typical for 1965–1968, and many such restrictions still exist. Changes since 1971 are discussed below.

Communist bloc. Neither old academicians attending congresses or conferences, nor the privileged tourist trips by members of the Writers' Union or trade-union elite, nor short-term exchanges of different delegations, scientific or scholarly, were able to fill the role of personal *free scientific cooperation,* available for British, German, Italian, or American scientists who can work abroad for long periods in the laboratories or technical research centers of their own choice on the basis of mutual interests. The superficial exchange which the official Russian bureaucracy considers "scientific cooperation" is not a substitute for the possibilities of foreign *training* or foreign *education* available to students of the Western world. The number of Soviet scientists temporarily working in the big research centers of Western Europe or in the United States and Canada is much smaller than the number of Polish or Hungarian experts working abroad, and the situation is changing slowly, if at all.

Party officials continue sticking to these restrictions because of the primitive fear of a possible increase in the "brain drain," defection, ideological "destabilization," or Western influence within the USSR. (A small increase in Western influence certainly would be an inevitable price for freedom of foreign travel. However, some increase is evident now anyway, especially among the young and teen-agers. This influence emerged without any travel abroad.) The "brain drain" is also not a serious danger, because the West now offers very few opportunities for professional employment. Many Western countries are at present suffering from a surplus of their own experts. Britain and Germany, for example, lost many thousands of scientists and experts through free emigration during the postwar period, a figure much greater than any number of possible "defectors" from the USSR. Nevertheless, the industrial and scientific development in these countries was able to afford this loss. The very freedom of scientific exchange makes these countries integrated parts of the world's science and technology anyway. For a British scientist working on the problems of nucleoprotein structure, it is now absolutely nonessential whoever makes the discovery relevant to his work, whether his own British colleague, or a French or Italian biochemist. However, a Russian biochemist living in Leningrad has a better possibility of collaborating with a colleague working in Vladivostok, some 5,000 miles away, than with a colleague in Finland, only eighty miles away.

Under such conditions the reorientation of Soviet science and technology away from the policy of duplication had a very small influence on the general rate of development of Soviet science. To find a unique problem for research is too difficult without *free* cooperation as the front line of scientific advance was still in the hands of the Americans and their West European colleagues. To prove that this wasteful delay in the modernization of Soviet science was artificial and mostly depends upon administrative and political restrictions, one can indicate that the situation was very different in the traditionally secret areas of scientific and technical research. The open cooperation between the USSR and the West was never possible in military fields, and independence had to be created here at any price. The price of Soviet achievements in these secret fields is higher, but the cost of development in strategic sciences never meant any serious trouble for the Soviet leaders.

RESTRICTIONS ON THE EXCHANGE OF INFORMATION

The new orientation for Soviet science and technology could bring Russian scientists to the front line of world science, but only when all the restrictions on the exchange of information are eliminated. However, the very system of political control over this exchange, together with the censorship and the delays in receiving even scientific information that is openly available in foreign academic and scholarly journals, put Soviet scientists in a rather difficult position. They could, in most cases, know last year's advances in their field, but not current developments.

The practice of copying foreign models which had been introduced by Khrushchev was not only politically motivated; the

Soviet Union simply did not yet have enough foreign currency reserves to buy the complex means of production. This shortage of hard foreign currency made it difficult to subsidize the subscription of the original foreign journals and the big industrial facilities that had been put into operation (approximately since 1954–1955) to copy these journals by photocopy methods. The number of such duplicated journals included more than 600 titles. This process was slow and the copies, in most cases, lagged behind the originals from six to ten months. All these problems of delays in the informational exchange have already been described in a special detailed study (9). I can only indicate that the situation which is described in this work for 1966–1968 was valid until 1974. Some changes which were made later on (the end of copying the journals, increase of direct subscription, and joining the Universal Copyright Convention) will be discussed in the next chapter.

However, the restriction in informational exchange does not only mean that the foreign information was not absolutely free and easily available. It also means that the Soviet scientists as well were not able to have a free exchange of information; publication of their own works in the international foreign journals was still very difficult. During Stalin's time, the scientist who, by some chance, was able to publish his work abroad could be accused of treason. Such "treason" in some cases was punishable not by prison or camp, but by a special "court of honor" and by public condemnation. Two "trials of honor" are well known, and were publicized and even made into a film in 1948. These were the "trials" of Professors N. G. Klueva and G. J. Roskin in 1946 for the publication of a small paper in one of the American cancer research journals (47), and the same kind of "trial" of Professor A. R. Zhebrak in 1947 for the publication two years earlier of a short article in *Science* (48) about Soviet biology, with a few critical comments on T. D. Lysenko. (The case of Klueva and Roskin was discussed at the Politburo level and became a sample of ideological vigilance for scientists.) During the Khrushchev period this practice was abandoned, but the very possibility of sending papers or research work abroad for publication still did not exist in practice. However, it was possible for prominent scientists to participate in some international congresses and conferences and through these channels to get their works published abroad. In 1968–1971 certain regulations for foreign publication of works in the international scientific press had been established, but the frustratingly slow and complex process by which submissions are considered for such publications could explain why so few scientists were ready to try. In 1971 no less than 98 percent of all research data received in Soviet laboratories had been reported initially in journals in the Russian language or in the languages of the other Soviet Republics (Ukrainian, Georgian, Armenian, etc.). Many of them had been translated later in full or in abstract form in different special translations or abstract services available in the United States and West European countries, but the time lag in this case was even more substantial. For the current extremely rapid wave of scientific and technological development, these methods of informational exchange were too slow and inefficient. Both sides of such exchanges were the losers. The damage, however, was more substantial for Soviet science because it was in a backward position.

In the social sciences—economics, history, philosophy, and others—the situation was much worse because the political divisions such as "Soviet," "Marxist," and "bourgeois" were still valid here. This made most of the foreign social and political literature as well as history, modern art, and even modern music unavailable for the majority of Soviet scholars and for the Soviet public at large. The foreign mass media, which are an important source of information for social scientists, were also unavailable. Some works and papers could be found in the special collections of large libraries and could be read by a few trusted professionals if special permission was granted for them, but most works were simply unknown or unavailable even in special collections. The rate of turnover of scientific information is now very high, the informational value of publications decreasing with the speed of 10–20 percent per annum and in some areas with an even higher speed of 30–40 percent per annum (this is easily determined by the "citation indexes," which show a rapid drop in the reference lists of works published in years past). Therefore the delays in the exchange of information made the qualitative gap for Soviet sciences unavoidable.

Self-Imposed Restrictions in Soviet Science and Technology: The Obsession with "Secrecy"

During Stalin's time, especially after the war, almost all branches of research had been considered secret or at least semi-secret. Secrecy was related not only to the economic or military value of information; in most cases the peculiar "priority consideration" was sufficient. Research in plant physiology or botany had to be considered as part of a special "Soviet" contribution to this branch of science, and it had to be safeguarded as "Soviet" up to the final results. (The bureaucracy had wrongly assumed that *any* research work had a well-defined beginning and final discovery which was the only major value of the research. This discovery had to be published by a Soviet author, who in this way became the owner of *priority,* the discovery therefore becoming a *Soviet* discovery.) It was also assumed that if somebody published the preliminary or intermediate parts of an experiment, foreign scientists would be able to continue and complete it, and in this way get the credit for the final discovery. (Klueva and Roskin were blamed because they published *preliminary* results, promising some new approaches in cancer research.) To escape such losses of "Soviet-made" discoveries in all the sciences, and the possibility that the USSR would not obtain the pride of priority, it was forbidden to exchange and publish *any* preliminary information or data about one's research work (even within the internal system of Soviet science). Any research was automatically considered classified until it was completed. To get a paper published in any area of science, it was necessary to have a special certificate signed by six experts that the work was complete, with all conclusions final, not preliminary. These certificates had to be submitted with the works which had been offered for publication. (Simultaneously with the submitting paper the duplicate manuscript had to be sent to the Special State Committee on Discoveries and Priorities and research workers could receive either the "certificate of priority" or "diploma of discovery.") These rules became serious obstacles for the publication of research works and very quickly they became a mere formality. The members of the publication commissions usually signed certificates without even reading the manuscripts. However, the research areas declared classified as a whole, where no results were permitted to be published, were so extensive that until 1953 practically no research using radioactive and stable isotopes could be published in the open scientific press. It did not matter whether these works were done with plants, bacteria, animals to study their metabolism, or on the problem of the chemistry of the separation of isotopes, which could be connected with military problems. Any research using radioactive substances was secret. This obsession with secrecy permeated chemistry, microbiology, and almost all branches of medical research, statistics, and agriculture. It reached a paranoid level, and from 1948 onward, even in school textbooks, no map could be published with the indications of meridians and parallels according to their normal pattern. Someone had decided that this would help possible enemies to find targets for military strikes, but because it would be too selective to eliminate the meridians and parallels from the Soviet Union and Eastern Europe only, the whole world temporarily lost them as well, although it was obvious to the experts that all the old maps, even the imperial Russian ones, would be quite sufficient for a possible enemy to find the geographical location of Moscow, Leningrad, or Sverdlovsk. (In a few years, the meridians and parallels were again permitted, but they were dislocated a little for textbooks, and "open" geography—correct maps necessary for ships' captains or airline pilots—is considered secret even now.)

The absolutely closed character of Soviet science began to change during Khrushchev's time. Soviet scientists at first started to receive (since 1955) standard reprint requests from abroad which before 1954 were simply intercepted and destroyed by postal censorship. Soviet scientific literature had been permitted to be sent abroad on the condition that this was officially printed literature for open circulation. Old restrictions were under reconsideration, and the very bold decision by Khrushchev to send a large Soviet delegation to the First United Nations Conference on the Peaceful Uses of Atomic Energy in Geneva in 1955 had a serious effect. The Soviet delegation consisted of many prominent scientists, academicians, corresponding members of academies, and professors. Several young scientists working at top-secret institutes on nuclear physics had been sent with their papers. As could be expected, the key figures of military-oriented secret research work were not allowed to travel abroad and even their names were not mentioned in the references. However, it was quite a surprise for the participants from other countries to see Soviet colleagues delivering papers and lectures about nuclear reactor problems, plasma projects, scientific and technological details about the first atomic power station, works on radiobiology, radiology, radioactive isotopes, and so on. Some research works were specially cleared and declassified for this international representation. The aim of this action was to show that Soviet science in these fields had reached the international level of development. The same clearance was given for many works of Soviet scientists for the First UNESCO Conference on the Use of Isotopes for Scientific Research (Paris, 1957) and for the UN's Second Geneva Conference in 1958. This decision made it possible to relax the rules on secrecy in many other branches of science as well.

Since that time, the exchange of scientific information with

foreign scientists has improved, and several modern research institutes, like the Institute of Nuclear Physics in Dubna, have
been transformed into *international* institutes. The personal exchange of scientists also started, but while these restricted moves
were certainly welcome and mutually useful, they were not
transformed into free cooperation. Many research fields considered open in almost all industrial countries were often within the
classified area in the Soviet Union. This was ridiculous for such
branches of technology as electronic and computer engineering,
where the Soviet Union was far behind the world level; however,
the secrecy often tried to hide the backwardness, not the advances.

If we consider the situation in 1965–1971, we find that the system of receiving clearance from a special commission of "nonsecrecy" and certification of the finalized character of the research
was still obligatory for any work before it was allowed to be submitted for publication. However, the number of members of the
commission was reduced from six to three. The editors of journals were simply not able to get censorship permission for the
publication of *any* paper, even for a review of foreign literature,
which was not accompanied by the certificate of nonsecrecy. The
censorship of all material for publication, functioning earlier
under the well-known name "Glavlit," was transformed into the
special "State Committee on the Protection of Military and State
Secrets in the Press." With departments in all parts of the country, it was responsible for the final approval of all open publications. The papers could be restricted from open publication not
only because they carried some military or state information, but
in many cases just because the information acquired by the research group might be used for one or another technical design
or process which could be licensed or patented. The lists of information considered forbidden for open publication were, and continue to be, too long to be discussed here. (The full extent of
these lists has never been disclosed.) However, the restrictions
can be easily judged from the experience of scientist colleagues

whose works were refused publication on the grounds of security risks or of the leakage of political, economic, or military information important to the state. I am not familiar with the situation in the fields of technology, chemistry, or physics. But in my own fields of knowledge—biology, genetics, and aging research—there were enough examples of "secret" information in regard not only to publication, but even to research. It is, for example, impossible to study and publish statistical information about the geographical pattern of age-related troubles and statistical evaluation of the causes of mortality (coronary and heart disease, pneumonia, cancer, etc.). In the textbooks on gerontology, this statistic is given for the United States, Britain, France, and so on, but not for the USSR. It is not permitted to publish statistical materials on endemic diseases, outbreaks of epidemics, figures of health conditions related to a profession or trade, mortality rates for different groups of workers (miners, workers in chemical plants, metal mechanics, seamen, etc.). The results of research into the geographical distribution of chromosomal abnormalities or hemoglobinopathies are not publishable in the USSR. Even such simple social-medical statistics as the number of pregnancies among girls of high (secondary) school age (under seventeen), the distribution of venereal diseases, the number of drug addicts, or even the consumption of alcohol per capita are among the state-guarded secrets.

Information about problems of pollution, aviation disasters, and other forms of catastrophe—figures easily available abroad—are considered classified within the USSR.

It is not only science which suffers from such restrictions—it is even more serious in social and historical areas. There is no time limit on the prohibition of access to the main state archives. Government and Party papers are closed for historians from the very beginning of the Soviet Union. Many names (Trotsky, Bukharin, even Khrushchev now) cannot be mentioned in papers on history without special permission.

All these restrictions in the vital fields of human life make serious research of many of these problems practically impossible. The secrecy of information is quickly being transformed into the *absence of reliable information*. The factor of secrecy has a great braking effect, because research workers very often stop the investigation as soon as they realize that the data they are getting will not be published. I know of several young geneticists, encouraged by the success of research of geographical distribution of genetic abnormalities made in other countries, who started similar work in the USSR, but soon gave up after failing to be allowed to publish their very interesting results. The same attitude is found in the applied sciences. Very few scientists and research workers are able to complete their work up to the final stages of new design, new invention, or new process. The successful end of research is often the coordinated result of the collective efforts of many groups. However, with the absence of worldwide, or at least nationwide, exchanges of information, such a synthetic, cooperative approach becomes too difficult and many works are left unfinished, unknown, forgotten, and finally lost.

DELAYS IN THE INTERNAL EXCHANGE OF INFORMATION: THE PROBLEM OF LOCAL LANGUAGES AND IGNORANCE OF INTERNATIONAL PUBLICATION FACILITIES

The road of scientific progress is full of small obstacles and problems which often seem insignificant, but which nevertheless are quite a visible deterrent. I have already mentioned that the rapid development of science and technology makes an open and efficient informational exchange extremely important. The reasons for the delays in the distribution of foreign journals were also discussed earlier. However, the delays in the distribution of internal open, cleared information were inexcusable. The figures published in Dobrov's book (39) relevant for 1967–1968 for Soviet technical journals, as well as my own estimations for biological sciences (9) made in 1969–1970, indicated that the average delays between submitting a paper for publication in one or another scientific journal and its actual publication vary from twelve to thirty months. There were no special Soviet journals in any field of research especially for accelerated publication, although there are several international journals of this type. The average publication time in the majority of European and American journals was, in 1967–1970, about six to eight months, but because this was already considered a slow rate, the proliferation of "rapid publication" journals for important results made the time intervals a matter of weeks, not months. Very few Soviet scientists were able to use these rapid-publication facilities of the international press, but with the elimination of some restrictions of Khrushchev's period, the names of Russian research workers became visible at times in such international journals as *Nature, FEBS-Letters, European Journal of Biochemistry,* and others. I do not think the same situation was true of technical, medical, or humanitarian journals, but in biology the move into the international press, however small, was evident. This tendency, as we will see, became stronger later. In 1969–1970, in biology, probably not more than a mere 1 percent of the entire output of information published by Russian authors was initially sent abroad for publication in foreign journals, and those were mainly in English.

Khrushchev's liberalization of internal national policy had a good influence on the development of local non-Russian national cultures. During 1956–1964 more films and plays were being made in the Soviet Union's national republics than during Stalin's time. More novels and poetry were written and published in the local languages—Ukrainian, Georgian, Armenian, Uzbek, and others. This was a welcome trend, as were the rehabilitation of several national minorities deported during Stalin's time from their homelands, and the end of openly hostile anti-Semitism. However, the situation in science is too specific to be decided by the same criteria. The improvement in the cultural development of smaller nations should not mean the use of local languages as a medium of scientific publication and communication. The encouragement of the use of local languages for scientific and technical publications was, in reality, a retrograde step—a move back from the general world trend.

The proceedings of republican academies—Georgian, Armenian, Ukrainian, and others—started to be published in the local languages, which made them of very little value. The majority of Russian scientists, and foreign scientists as well, could not read these papers in the original, and they have simply been lost to the mainstream of science and technology. It was not unusual for the research institute in Estonia to receive an official letter written in the Uzbek language from the Uzbekistan Soviet Republic, and it could take months before anyone could be found who was able to read it. In retaliation, the reply would be written in Estonian, and this would create the same situation in Uzbekistan. (A special multilingual translation service has been established in Moscow to deal with such cross-misunderstandings.) After a few years of frustration, the Russian language again was made the official language for internal communication. Even then some purely scientific journals continued to be published in the local languages. Scientific books printed in Georgian or Armenian, as a rule, had a very small circulation and sale; perhaps not more than a few dozen experts would be interested in buying such books. The major part of printed academic materials are left unsold and usually pulped down within a few years. This way of satisfying the feelings of national autonomy was probably good as a political gesture, but was a deterrent in science.

In 1965–1971, the situation was partly improved, and as regards science this can hardly be considered a forced "Russificiation." In many cases the "Russification" of science raised the quality of research, as well as the quality of education, in the Soviet national republics. I have expressed this view several times and have been accused of being a "Russian chauvinist." Certainly local language is advantageous for the development of local literature and poetry, folklore and culture, as well as all activities of national origin. But for science, which is international, the local language is a poor way of communicating. I am not in favor of a compulsory use of the Russian language when the paper is written by a Soviet citizen on protein biosynthesis, or theoretical physics. It would be better for the international exchange of knowledge if it was published, say, in German or English. Many academies of sciences—for example, in Poland, Hungary, and Romania—choose English as the language for advanced scientific journals. It is also necessary for the Soviet Union to publish some internal journals in English as well, but so far this has not yet been established. Not only is it important for the outside world to be better informed about research work within the Soviet Union; it is even more important for Soviet science to use the common language of communication.

Repressive Measures against Dissident Scientists

The majority of scientists in every country are not politically oriented. Generous financial support from governments can usually create any kind of research, good or malicious, for peace or war. Political systems are influential, but in the end scientists, not politicians, create the means of self-destruction for the human race as a whole. The consciousness of this responsibility has now emerged, but it is still too weak a trend in science, and it is also too late. The stockpiles of nuclear, chemical, and biological weaponry have already reached super-dangerous levels, and it was the scientists who discovered and designed these means of mass destruction. Many scientists talk about humanism, morality, and justice as the main principles of international relations, but we all know that the scientific progress of the last few decades (or the so-called Scientific-Technical Revolution) was too closely interrelated with military objectives. Space research, nuclear physics, and work with radioactive labels which accelerated almost all discoveries in biochemistry, chemistry, and many other fields of *pure* science had been just by-products of military programs. Politics plays too important a role in shaping scientific research in every country and especially in technology. This is the case in all the big countries, and the maltreatment of "dissidents" in science and technology is not exclusively a "Soviet" phenomenon—only the methods are different. The notorious case of J. Robert Oppenheimer in the United States during the cold war is probably not comparable with the "trial of honor" of Professor A. R. Zhebrack, the death sentence for academician Nikolai Vavilov, or the more recent campaign against academician Andrei Sakharov, but it was a shameful episode in American life. In all countries during the last century, either "free" or totalitarian, the state (or ruling power groups) decided upon the direction of scientific technological development, either by selective distribution of financial support or by utilizing more direct political pressures. In many cases, research work is a branch of the military service, and when a scientist signs the declaration necessary for clearance for secret information (the obligatory step for at least half of all senior research workers in the USSR), this is nothing other than an oath of allegiance and enlistment for military service, though in "plain clothes." The privileges which scientists receive from the state (and readily accept), the financial support, the many state honors and prizes, make the ruling political groups expect that they will also provide the necessary level of obedience. At the same time, the research workers, for the sake of high productivity and success in research, need much more intellectual freedom and political rights than other classes and groups of society. Poor peasants with large families or coal miners in remote areas of the Siberian coal fields are not at this time concerned about freedom of the press, freedom of informational exchange, freedom of foreign travel, or freedom of emigration. None of these rights are necessary for their hard work. For true scientists the basic human rights are inseparable from their right of research. Violation of these rights only damages the creativity of work. In the Soviet Union, the top scientists had often exercised practically unlimited freedom and had been in fact members of the ruling group without formal membership in the Politburo or in the government. Igor Kurchatov, in nuclear industry, and Trofim Lysenko, in agriculture, during the postwar decade had much more power than ministers in these fields. Both Kurchatov and Lysenko were able to force some ministers to resign if they found them inefficient in the management of the "state-important" scientific programs. Even now, the position of the president of the Academy of Sciences of the USSR is more influential than the position of minister of the government, and the change of the Academy president is a more serious affair for Party leadership than the change of a minister in most branches of industry.

This especially privileged position of prominent scientists made the Soviet government and Party leadership almost totally unprepared when, in 1966–1967, the scientific elite started to express some dissent in political views and took some definite political action. This action was still rather modest—mostly letters to the Party Central Committee or Party Congress (in 1966), with loyal, mild, but quite definite protests against some political actions, against attempts to rehabilitate Stalin and to restore political trials (like that of Sinyavsky and Daniel), to support dissident writers demanding the abolition of censorship, and some others. This kind of dissent could probably be expected from young scientists and intellectuals, quite a few of whom tried to criticize the regime and had been arrested in 1966 and 1967. (The first troubles of the young intellectuals A. Amalrik, V. Bukovsky, A. Ginsburg, Yu. Galanskov, and others, unknown in 1966 but later widely publicized, started just at this time.) These arrests and poorly conducted trials induced even stronger dissent among prominent figures in science, technology, and art. The first "Open Letters," with hundreds of signatures, which were later published abroad, became the signs of an organized movement for democracy in 1967 and early 1968. This movement was concentrated mostly in research and university centers, and was greatly influenced by the developments of the "Prague Spring" in

Czechoslovakia, where the democratic measures were not just discussed but really introduced by the Central Committee of the Communist Party. The Czechoslovak model of democratic socialism had a hopeful and attentive following among the majority of Soviet intellectuals. It was, however, also under strong and hateful pressure from dogmatic and hard-line leaders of the USSR, East Germany, Bulgaria, Hungary, Romania, and Poland. Not all the leaders in these countries followed the Czechoslovak development with the same feelings, and not all the leaders in the USSR had the same attitude toward this problem. It was such a split that made the situation even more critical.

The decision to take some active measures against internal dissent had been made in the USSR before the tragic events of August, 1968. These measures were not yet strongly activated. Party and administrative actions were used first. The list of signatories of various protests with full names and positions had been available, and some of the documents had hundreds of names. The situation was reminiscent of the famous Mao slogan, "Let a hundred flowers blossom and a hundred schools of thought contend." This campaign also induced a wave of criticism against the regime by the Chinese intellectuals. However, when Mao and his associates started to feel the anti-establishment influence of this tolerance and the truth in the bitter criticism, they sent thousands of critics to "thought reform" labor camps. The names of critics of Soviet policy have merely been exposed to officials during a short honeymoon of "Chinese style" freedom.

There was always some suspicion when the authors or initiators of insignificant documents or badly worded protests tried to get as many signatures as possible and approached famous and prominent people with the request to sign (without proper warning of intended foreign publication and all the risk involved). It was difficult to be sure, but my own opinion at that time was that some long lists were not necessary, and that very many of those who signed such appeals to the Central Committee would not have done so if they had known beforehand that those appeals were to be sent not only to the Central Committee but to the foreign press as well. Just these cases of *foreign publications* with the lists including these names were used in 1968 by the KGB and Party organizations for administrative and party measures. Those who were listed in the *New York Times* or the *Daily Telegraph* or in the émigré press as signatories were called to the meetings of Party organizations where they worked. They faced the Party activists or administration to explain how and why their names were used by the anti-Soviet or émigré press. Those who confessed to having been misled and having made errors and who repented their actions were often forgiven after being reprimanded. The reluctant ones were usually expelled from the Party or dismissed from responsible positions, and sometimes they were fired from their institutes. Scientists who had been classified for secret work were usually stripped of those clearances, and in that way lost their jobs automatically. These measures were taken against junior as well as senior scientists, and also included very prominent figures. Academician Andrei Sakharov lost his clearance and was not able to enter his institute (which was near his apartment) at the end of 1968.

Dissident writers were blacklisted through censorship, and they suddenly found that there was no room in journals and magazines to publish their stories or essays. Even the proofs were often destroyed, the usual excuse being the "shortage of printing paper."

Young scientists awaiting final approval of their theses in the Highest Qualification Commission in Moscow soon found that they waited endlessly (some waited several years), and some successfully defended works were turned down.

The situation became even worse after the tragic events in Czechoslovakia. Meetings were called at all the institutes, plants, collective farms, and other establishments to approve the occupation as an act of international solidarity and help. Those who refused to attend were expelled from Party membership and lost their administrative or responsible positions. Experts and scientists were transferred to lower positions or even lost their jobs. Political considerations certainly prevailed under those of scientific developments. Many excellent scientists were fired from professorship positions and lost their laboratories or departments. I could easily see these trends in Obninsk, where I lived. The whole theoretical department of the Institute of Nuclear Energy in Obninsk had been liquidated because the best men had been fired. In the new Institute of Medical Radiology in Obninsk, the Department of Genetics had been closed in spite of its absolute necessity for radiobiological work, and N. V. Timofeev-Resovsky, the head of this department, was sent into retirement. My own new Laboratory of Molecular Radiobiology was closed, and I lost my job at the institute. Some scientists had been fired from the nearby Institute of Radiochemistry. Open dissent in the scientific community was reduced, and protest letters with hundreds of names never appeared again.

Science and research suffered much more. The cost of repression for Soviet science is difficult to evaluate. The loss certainly was not as dramatic as it had been in 1937–1938 or in 1948–1953. For the many dismissed senior scientists or professors in 1968–1969, there were enough new candidates to fill the vacancies, but science with its lost freedom could never be good enough again. When political loyalty is ranked higher than creative talents and abilities, the whole scientific establishment is not well enough qualified to compete with the advanced international level.

One of the indirect methods of political control of science was a certain program for reconsideration of Khrushchev's idea of the small, quiet academic satellite town, like Dubna, Academgorodok near Novosibirsk, Obninsk, and some others. These towns with a higher standard of living and a concentration of intellectuals, and active student youth, became in 1967–1968 certain centers of intellectual dissent. Activists of the "democratic movement" always received a good audience here, and many "summer" or "winter" schools of advancement in science organized in such towns had been schools of political discussion as well.

After 1969–1970, these attractive towns started to lose their "science campus" appearance, and planning orgaizations located many new industrial developments in their place. Academgorodok in Siberia was simply joined administratively to the neighboring big industrial city of Novosibirsk and in this way lost not only autonomy but also its individual name. In towns which were situated at further distances from major cities (e.g., Obninsk or Dubna), large industrial plants were built and the towns became cities with the working class as the major part of their population. (In Obninsk, the population grew from 30,000 in 1968 to 120,000 in 1974.) The locational isolation of troublesome scientific and intellectual communities had been ended.

The repressive measures against dissidents induced a strong new movement—the demands of many Jewish intellectuals for emigration. The Jewish community in Soviet science, technology, and the arts had always been substantial. Many of them suddenly revolted openly against repression and anti-Semitism. This protest inevitably took the form of a trend toward emigration. This was a new problem, and again, as in 1966–1967 with the general dissent, the government was not ready and simply did not know how to deal with this phenomenon. Hundreds of Jews were dismissed from their jobs, some accused and sent to the camps. The shameful "educational tax" was introduced for emigration, but this reaction only backfired. Contradictions within the scientific community reached a rather high tension and all the new repressive measures, including the most serious and terrible ones—psychiatric clinics and hospitals—did not really help to suppress dissent. They only added enormously to the bad reputation of the Soviet leaders and damaged the Communists' international image. Neither new space experiments nor the restrictive and selective permission for Jewish emigration was able to impress public opinion abroad. More radical changes in internal and international policy became inevitable, as did a move away from wasteful duplication of Western technology and experience toward a more integrated economic and technological policy in line with the rest of the industrial world. In the middle of 1971, this new policy began to emerge after the Twenty-fourth Party Congress in Moscow. It took several years for the outside world to consider this new policy seriously and to discuss it as the background for reassessment of the relations among East and West. Neither the Russians nor the Americans, the two main superpowers, were able to explain correctly the substance of change, but the change was obvious. I indicated earlier that it had received the name *détente,* and the most important question had been asked everywhere—is the *détente* policy real and sincere, or is it a political game? Was it really necessary for Russia to cooperate with the West, or was it just a way to obtain some concessions from the West without any particular changes of real importance within the Soviet Union itself?

Chapter 8

Détente and Soviet Science (1972–1977)

Détente had been started by the USSR in 1971 for many reasons, mainly strategic and military; the arms race had reached the point where the two superpowers realized that they were equally capable of destroying the world, including themselves. The United States could still claim superiority in some military areas—for example, aircraft carriers and electronics, research on "cruise" missiles, etc.—but it was counterbalanced by the Russian lead in "super" missiles, rockets with a payload of 16,000 pounds, and in developing "mobile" missiles. Nuclear arsenals were also comparable.

The Soviet Union certainly led in conventional weapons such as tanks, guns, antiaircraft rockets, and other military equipment which had not lost its importance, either strategic or tactical. As the arms race became more competitive, the speed of accumulation of nuclear and other arms was high in both countries, but it was higher in the USSR. Comparative equality between the two powers was reached at the price of wasteful overproduction, and it was clear to both that they must somehow stop this trend. It was also clear that the USSR, which had been much weaker ten years ago after suffering some kind of military humiliation during the Cuban crisis in 1962, would be the first to offer the *détente* policy as soon as it had overcome its inferiority complex. There were too many politicians in the United States and Europe, as well as many dissidents in the USSR (A. Sakharov, A. Solzhenitsyn, I. Shafarevich, A. Amalrik were the best known among this group), who, misled by their already obsolete argument of the recent comparative weakness of the USSR, tried to warn the West against the *détente* policy, and declared that *détente* between the more powerful Western world and the weaker, more backward eastern bloc would give Russia the means of catching up technologically in the militarily important areas, without giving any advantage to the West itself. To issue such warnings from within the USSR takes great courage and determination as well as strong feelings against the Soviet regime. Such warnings declared openly and with the maximum publicity were associated with great personal risk. We can admire the courage of such people, but it is not my task here to describe the arguments of the intellectuals who tried to warn the West about the danger of *détente* or who argued that *détente* would be mutually advantageous.*

Since the general dispute on various issues of *détente* has already been adequately covered, it is unnecessary for me to analyze here the arguments expressed by many scientists and intellectuals during this discussion, which still continues and certainly will continue for many years to come. My main task here is to describe *what the* détente *policy and the new climate in*

---

* The whole controversy of opinions and actions about the advantages and dangers of the *détente* policy has been reflected accurately in many recent pamphlets and more comprehensive works, which cover the discussions of 1972–1973 and later disputes (49–55). My own position in regard to this problem was made clear in my testimony before the Committee on Foreign Relations of the United States Senate during its hearings on *détente* in 1974 (56).

international relations really meant for Soviet science and technology.

A cautious approach toward *détente* had been started by the USSR during Khrushchev's time. It was an inevitable result of the acknowledgment of Soviet backwardness in many fields of science and technology. The aim of Khrushchev's peaceful trend of coexistence also was clear—to make Russia a more modern, equal, and prosperous country by way of assimilation and duplication of foreign technology and experience. In 1972 living standards and the production of consumer goods in the USSR were below those in the United States and Western Europe, but nevertheless the new *détente* approach had quite a different background. It was not induced by backwardness in science and technology in the USSR, as many critics of *détente* thought to be the case. *It was induced by the feeling of military equality, by the success of military science and military technology,* and indeed by the feeling of some superiority in the development of conventional arms. A highly centralized socialist economy is better placed for awarding high priority to the preferential development of selected branches of science and technology. The government can manipulate the economy much more easily than can the governments of democratic countries. Neither the electorate, the press, the opposition, nor the danger of strikes or parliamentary crises can influence the distribution of national resources. At the time when these resources were not so well developed, the USSR could not reach the American level, even in the selectively sponsored fields of science and industry. However, in 1972 the *heavy industry* of the USSR had caught up with the United States and this was the crucial point. Some economic reforms, implemented since 1965, had removed many obstacles which had previously retarded technological progress.

From that time onward the Soviet military machine was able to develop more rapidly than that of the Western world. This acknowledged higher quality of some branches of military technology in the United States was no longer a decisive factor, and its qualitative superiority in too many other branches was no longer certain. The military defeats of the Western "client" regimes in South Vietnam, Angola, Mozambique, and elsewhere, the deadlock in the Middle Eastern military conflict in 1973, did more than compensate for the humiliation of Soviet military power during the Cuban missile crisis in 1962. The offer of *détente* in 1972 was not made because of weakness, as it was in 1955 when Khrushchev frankly wanted to reduce military spending for the sake of developing industry and agriculture. The West did not trust Khrushchev's approach, and the contradictory and unpredictable character of Khrushchev's actions made the reluctance of American and other Western leaders understandable. The cold war could not end overnight, and the over-all superiority of the West seemed to be the best guarantee for peace.

But over-all American superiority no longer exists. Western Europe is very vulnerable and unsafe, and it also has internal problems. *Détente* now is mutually advantageous, and in the short run it could be even more important for the West than for the Soviet bloc. The situation is changing so quickly that those Russian dissidents who tried to warn the West against *détente* two or three years ago now talk in despair about the "defeat" of the West, the "retreat" of the West, the "weakness" of the West, the "selfishness" of the West, or even about its "capitulation." The debates on this issue are now so wide-ranging that in this book I decided to pick out the problems of militarily oriented science in the USSR for special consideration. It is interesting to see why and how military technology could become so mighty and all-powerful in a country which is not yet able to solve so many vital economic problems.

The Possibilities for Selective and Independent Development of Militarily Oriented Sciences and Technology in the USSR

Genetics and biochemistry, like other natural sciences, are open and without frontiers. Attempts to confine them within state frontiers could have negative consequences for all nations.

By contrast, military sciences and technologies were almost always secret and not shared with other countries. Worldwide co-operation in these fields would be nonsense. Where such cooperation becomes possible, military hardware products will no longer be necessary. The main objects of any militarily oriented sciences are to discover, first, the knowledge of how to destroy any adversary power the most efficiently and with minimal risk of self-destruction and, secondly, creation of the means to this end.

Before the outbreak of the war with Germany in 1941, the importance of military technology was underestimated in the USSR, and the notion that war against the Soviet socialist proletarian state would induce internal revolt, uprisings, and strikes of workers in the aggressor countries was still the official belief. "International solidarity of the proletariat" was considered the real force which would compensate for the isolated position of the Soviet Union. This myth, coupled with Stalin's terror during the 1930s which destroyed the Soviet technological elite, gave the German army technical superiority at the beginning of the war, and many millions of Soviet soldiers and civilians paid the price with their lives in order to save the country from the mistakes and brutality of Stalin's rule. Help from the outside did come during the war as well. It was, however, not the help of the "international proletariat," but military and technical assistance from the advanced capitalist countries, the United States and Britain, which were involved in the war themselves. France, Belgium, Holland, Poland, and other European countries were already under German occupation, and Britain would certainly have been the next to fall if Russia were defeated too quickly. Therefore, the joint efforts of the Allies were not dictated by humanitarian or social cooperation but by the sheer strategic necessity for winning the war.

Soviet military technology developed quickly and dramatically during the very short period after the beginning of the war, and by 1943 had overtaken the earlier German technical superiority in many important fields. However, the United States, which did not suffer occupation, the destruction of its industry, or the bombardment of its cities, ended the war as the strongest world power, in both the scientific and the technical fields, with atomic bombs whose terrible power had already been demonstrated in Japan. The message of Hiroshima and Nagasaki was well understood in Russia, but in quite a different way from that which the United States might have expected. In 1926–1927, when the Soviet Union programmed industrialization, the task had been absolutely clear—to make Russia the most modern industrial power in the shortest possible time *at any price,* this being the only possible means of survival. Human life was not important. Living conditions were ignored, and new industrial complexes and plants were built without either flats or houses for workers. They lived either in special tents or in temporary wooden barracks with four, six, or even ten beds packed into each room, such barracks surviving in many industrial areas until 1956–1957. The situation after the war made the development of military industry and military science even more urgent, with three prime objectives: nuclear power, supersonic military aircraft, and rocketry.

Almost half the country had been laid waste by the war. About twenty million people, mostly young, had been killed. More than ten million were left disabled. The labor prison camps were overcrowded as never before—with about five million German prisoners of war, several million repatriated Russian prisoners of war, civilians considered "collaborators," and many others. By 1945–1946 the population of the camps had trebled. (The number of labor prisoners during the war was sharply reduced because of the high mortality rate among prewar detainees.)

Agricultural production was only 60 percent of the level of 1940, and there was no means of raising this level quickly. Nevertheless, it was science that received priority for development, with top priority being given to physics and missile and aircraft technologies.

A totalitarian country has many means of improving the performance in certain specific fields when competition from other fields is limited by concentrating available resources on some special projects. When the United States decided to use *two* atomic bombs during the final stages of Japan's resistance, American military experts certainly counted on not less than ten to fifteen years of atomic monopoly, so important for political dominance. They did not realize, however, the capability of a totalitarian state to concentrate its efforts on any particular development. Totalitarian states with a capitalist economy (like Hitler's Germany) are not able to reprogram industry and science with the efficiency and speed of a socialist totalitarian country where all the means of production, education, and research are in the state's control. Democratic "open" societies like those of Britain or the United States are also much more vulnerable to foreign intelligence penetration than a totalitarian country of the Soviet postwar type, where permanent surveillance through overt and secret informers was a way of life for the whole population, and where all free channels of communication with the outside world were closed.

Intellectual talents are not given to everyone, and special ability is as important for science as for any other kind of human activity. The United States had accumulated a unique pool of talent both before and after the war, owing partly to large-scale immigration from Europe. But science does not necessarily attract such talent in the United States or in Western Europe. Many nonscientific businesses and professions, which offer higher pay and more rapid advancement, also attract young talent. In the Soviet Union's planned and regimented life most of these alternative opportunities do not exist. Science is therefore not only the best-paid field of work, but it also offers the most freedom and advantages. Party and government administrative posts carry higher privileges, but they are not open to all, they are not too safe, and anyway they are not for young people. Such positions probably attract those with a special taste for power but not graduates from universities. From this point of view, *human resources* available for research work are greater in the Soviet

Union than in many other countries. A scientific career would rate very high in any questionnaire of preferred professions if such a questionnaire could be given to the most able students. Unfortunately, such inquiries and research are not carried out in the Soviet Union, and educational sociology is a very new branch of research. Such questionnaires are important because the career decisions of able students and applicants for university education and other kinds of higher education determine to some extent the future distribution of talent among different fields of human activity.

In the United States, where such questionnaires have been distributed, the picture of career choice is very different from that in the USSR. Nichols (57) reports that in a group representing approximately 1 percent of high school seniors who ranked highest in scholastic aptitude, 28 percent gave "scientific research" as the career choice of first preference. In the USSR the most able 1 percent would certainly give "scientific research" in more than 50 percent of cases, not only because of the popularity of science, but also because the work is much better paid than the work of doctors, lawyers, engineers, or teachers.

In the USSR the change of a profession during the course of one's education is rare. Because of the separation of professional groups in the Soviet educational system, it is difficult to change direction, for example from agriculture to medicine or economics, during college life.

The popularity of one or another reasearch activity is not a factor in the distribution of research workers. Areas of research are determined by the state, and such fields as physics and technology receive the majority of all graduates for "fixed vacancies." In 1974 the number of vacant postgraudate research positions was three times higher for physics than for biology; for technology, ten times higher. Military research, of course, receives the lion's share of research aspirants. More than half of the research workers in physics and technology work in areas related either directly or indirectly to the military.

Photographs

Peter M. Zhukovsky

Peter M. Zhukovsky, professor of botany at the K. Timiriazev Moscow Agricultural Academy, was my first teacher and protector. I started working during the evenings in his department as soon as I became a first-year undergraduate student at the Academy in 1945. (The name "Academy" was historical; established as an educational institution in 1865, the Academy was the first higher school of agriculture in Russia.) In early 1948, while I was still an undergraduate student, Professor Zhukovsky secured my further career as a scientist by making me the co-author of some of his publications. He knew that politics would make it difficult for me to get a postgraduate position (my father was arrested during Stalin's purges and died in the Kolyma labor camp in 1941), and he encouraged me to write a doctoral thesis while I was still an undergraduate. I received my Ph.D. in 1950.

Peter Zhukovsky began his post-war dispute with T. D. Lysenko in 1946. Many historians of science feel that Zhukovsky disgraced himself in August of 1948 when, under strong pressure, he publicly "confessed" his "errors" and promised loyalty and support for "Michurin-Lysenko" biology.

During this time I was at the Crimean Botanical Garden on the Black Sea coast. Immediately after the ill-famed Moscow conference where Stalin approved Lysenko's main report and forced Zhukovsky to "repent," Zhukovsky left Moscow to spend two months among the plants in the Crimean Botanical Garden.

"I made Brest peace with Lysenko"—these were Zhukovsky's first words to me when we met in the garden, and he looked around afraid that somebody else could hear. His sentence explained everything to me. The Brest peace treaty was a treaty which Lenin had insisted on signing with the victorious Germans after the Revolution. Lenin had explained that the treaty, though shameful, was temporary; that Germany would lose the war and Russia would be able to retrieve most of the land it had surrendered to the German Empire—Poland, the Ukraine, the Baltic lands, and many others. Lenin's tactic proved partly right.

Zhukovsky's similar tactic also produced some results. His false repentance enabled him to preserve his position and his pupils, and to support them during difficult times. In 1951, while Stalin was still alive, Zhukovsky resumed his struggle against Lysenko. Later he was appointed director of Leningrad's All-Union Institute of Plant Breeding (an institute created by Nikolai Vavilov), but was dismissed from this position in 1961 when Lysenko again became president of the Lenin Academy of Agricultural Sciences.

After Lysenko's fall in 1965, the first Soviet journal on genetics was established, and Zhukovsky became its first editor-in-chief. He led the journal successfully until 1975, when he died from a heart attack at the age of 86. At the time of his death Zhukovsky was sitting in his study working on his eleventh or twelfth major book.

Zhukovsky was an extremely productive scientist whose main contribution to the agricultural field was his discovery, in 1924–25, of a unique species of wheat, *Triticum timofeevi zhukovsky,* which was immune to almost all wheat diseases and parasites. This wheat became the main source of genes of immunity for all subsequent work on wheat hybridization.

Vsevolod M. Klechkovsky

Vsevolod Klechkovsky was a professor of agrochemistry and biochemistry at the K. Timiriazev Moscow Agricultural Academy. I began working in his department in 1951, first as a junior, and later senior, research scientist. Professor Klechkovsky pioneered the use of radioactive isotopes in Soviet agricultural and plant chemistry research. He had many research interests and published works on agrochemistry, biochemistry, biophysics, and statistics. He also produced a book and approximately twenty papers on theoretical chemistry and theoretical physico-chemistry.

Klechkovsky recommended that my research of 1956 be included in the International UNESCO Conference on the Use of Isotopes in Scientific Research. My 1957 trip to Paris for this conference, my first trip abroad, seriously influenced my attitudes toward many problems. Professor Klechkovsky was also a member of the Soviet delegation at the conference. During this time Russians abroad always lived in double rooms, and it was recommended that they walk around in pairs. Klechkovsky and I were almost always together because of this rule.

Klechkovsky had been involved in a serious dispute with Lysenko since 1951, when Lysenko developed a unique pseudo-scientific theory of plant nutrition and started to make recommendations about fertilizers. In one of his many speeches on agriculture, Nikita Khrushchev strongly blamed Professor Klechkovsky for organizing opposition to Lysenko.

When the first version of my book about Lysenko became known in *samizdat* in 1962, Professor Klechkovsky refused to follow the Party Committee's recommendation to dismiss me from my position and send me to the Academy's experimental station in the Tambov region. He was one of the few prominent figures at the Academy who strongly defended me at the "special" meetings, thereby giving me the time to find a new and even better position at one of the research institutes of the Academy of Medical Sciences. Klechkovsky wrote me an excellent recommendation and convinced the department's party organizer to sign it with him. In 1963 both men were reprimanded for this "protectionism" and liberalism.

Vsevolod Klechkovsky died from a heart attack in 1971 at the age of 72. He had just begun the organization of a new Research Institute of Agricultural Radiobiology.

Boris L. Astaurov

Though I did not work with Boris Astaurov as a research scientist, he helped me greatly in writing my book on the history of the genetic controversy, which was eventually published in the United States under the title *The Rise and Fall of T. D. Lysenko* (1969).

I first met Astaurov in 1957, when his work as a geneticist was practically underground, carried out with some cytological cover research. During the twenties Astaurov had worked with the founder of Russian genetics, S. S. Chetverikov, and during the thirties he worked with Professor N. Koltsov.

Boris Astaurov was a man of great wisdom and courage who was willing to risk everything for scientific principles. When the Genetic Society of the USSR was created in 1966, he was elected its first president. In 1967 he organized the Institute of Developmental Biology. Astaurov tried to get my book on the history of Soviet genetics published in 1965, and convinced the Academy of Sciences Presidium to create a commission to consider its publication. The commission consisted of twelve prominent scientists, and it made an unanimous decision to publish my book. However, the decision was vetoed in 1967 by M. Keldysh, president of the Academy of Sciences of the USSR, an action which led to my decision to send the manuscript abroad for publication. After Lysenko's fall, Astaurov was the first to attempt the revival of the study of human genetics in the Soviet Union.

When I was arrested and sent to a mental hospital in Kaluga in 1970, Boris Astaurov was the first person to visit me, the day after my arrest. When Astaurov told my wife and me that he would do everything to obtain my release, I knew that he meant it.

Boris Astaurov died from a heart attack in 1974 at the age of seventy.

Igor Kurchatov and Peter Kapitsa

A. S. Serebrovsky

Andrei Sakharov

N. I. Vavilov

Lev Landau

T. D. Lysenko

S. P. Korolev (left), I. V. Kurchatov, and M. V. Keldysh

N. K. Koltsov

N. V. Timofeev-Resovsky

A. R. Zhebrak

G. D. Karpetchenko

S. S. Chetverikov

V. A. Engelgardt

V. N. Sukachev

N. N. Semenov

For many reasons research in the military area is far less backward than Soviet science in so many other fields. I will try to indicate these reasons without, however, judging their comparative importance.

1. The military research establishments receive the best of the general pool of young talent. The rate of competition at the universities and higher technical schools is greatest for physics, electronics, aircraft engineering, and so on. For each fixed vacancy in these fields, there could be as many as ten or twelve applicants (compared with two or three for educational vacancies in medicine, agriculture, chemistry, etc.). The most important research centers in military technology and nuclear physics have first priority of selection among graduates and postgraduates, and they certainly try to employ the best. Soviet graduates and postgraduates are not free to accept *any* employment—they are under legal obligation to accept one of the positions which are offered to them by the state (through so-called "distributional commissions"), and to work a minimum term of three years *without the right to resign.* To leave a state-appointed position during this three-year period is considered a crime, and nobody has a right to employ such "deserters" freely. When the obligatory three years are over, the young expert is more or less free to look for another job, although very few who work in classified fields of research really want to do so. The second point may help to explain why.

2. Most professional research positions connected with classified work receive 20 percent or more salary than similar positions in "open" research institutes. The institutes connected with classified work usually have first call on the best equipment, funding, foreign currency grants, literature, and even living accommodation—necessities for successful research work.

3. I earlier advanced some arguments about the depressive influence of secrecy on the quality of research because of the lack of free exchange of information. This is true of many fields of research, like radiobiology, medical statistics, and others, which are "classified" in the USSR, but not classified or secret in many

other countries. From this point of view, Soviet science needs many more "open" fields of research than at present. However, there are several secret projects and areas of research which are secret everywhere. Nobody makes a free exchange of information in military aircraft technology or on work related to nuclear weapons, missiles, or conventional military equipment. In these areas, the winner is the country which has the better experts, better financial support, intelligence service, and organization, and stronger pressure from the state to reach certain standards and targets. I do not know how much all these conditions influence, for example, American work in similar areas, but in the USSR the state encourages them to be as strong as possible. That is why the Soviet Union made the atomic bomb so much earlier than American experts or politicians could have predicted. That is also why the hydrogen bomb was developed even earlier in the USSR than in the United States. Probably the same explanation can be given to explain why the fifteen-year-old military "Foxbat" (MIG-25), brought by a defecting pilot to Japan in 1976, caused such excitement to all foreign intelligence services. The plane that landed in Japan happened to be only the prototype of the MIG-25. Nevertheless, it was regarded as a gold mine for intelligence. In speed and altitude capability, the new models of the MIG-25 (as well as the older models of the 1960s) far outclass any comparable American and Western aircraft.

4. The situation in the crucial areas of military research was about the same in peacetime as it was during the war—superiority over the enemy had to be reached *at any price*. Many Soviet military victories would have cost much less in terms of the lives of Soviet soldiers if the strategy of military command had been more flexible. More than a million Soviet soldiers were killed during the last two weeks of the war for the sake of being the first in Berlin—a priority which was unnecessary for final victory, which was about to be achieved anyway. It is clear from Marshal G. K. Zhukhov's memoirs (59) that although the final costly advance may have accelerated the collapse of the German

army, the end was only a matter of days away. Prestige and politics rather than military considerations were involved in many of the battles of April and May, 1945. During the last few weeks of fighting, the Soviet army lost more soldiers and officers than the British and American armies lost during the whole war.

The same attitude was transferred to science when the state sought to gain military advantages. The story of American research into atomic weaponry is well known, as are the efforts which were made to reach certain objectives. The story of Russian work in the same field is not yet openly told. Two books about the late Igor Kurchatov (28, 29) are the main sources, but they give only a small part of the picture. Even these sources make it quite clear that Kurchatov, who was scientific director and organizer of the whole atomic project, had incomparably more independence and administrative, political, and scientific power than J. Robert Oppenheimer, scientific leader of the A-bomb project in the United States, and Edward Teller, research director of the H-bomb project. The first thermonuclear explosion by the United States was in 1952, but this was not a bomb which could be delivered by plane or rocket. The explosion of a thermonuclear device of military design took place in the USSR on September 12, 1953, about six months earlier than in the United States (March, 1954). This advance in nuclear weaponry was an impressive scientific military achievement, and it was possible because Kurchatov had dictatorial power, a kind of "marshal" power over everything necessary for the task. After 1949 several other scientists were given similar authority over their spheres in the thermonuclear project. Thus ministers of all branches of industry were under an obligation to provide everything immediately whenever asked by the research team. (In the Politburo of the Party Central Committee, Beria was responsible for the backup to this project, and under his direction many large construction projects were carried out with unlimited use of prison slave labor.) The same acceleration applied to other top-priority military projects during the post-Stalin era, but without the use of the labor camp work force. This rapid development probably cost much more than similar projects in other countries, in part because the final product could be less sophisticated and have fewer different electronic devices than the U.S. models. Nevertheless, from the most basic personnel weapon, like the famous Kalashnikov assault automatic rifle, to modern tanks, antiaircraft missiles, fighter planes, and thermonuclear weapons, Russian military science and technology have proved themselves independent and highly effective. The rate of development and the rate of buildup of their products have been more rapid than in the United States during the last decade, so that the accumulated power, not the weakness, of Soviet military capabilities made it inevitable to consider seriously the offer of *détente* by the USSR.

5. The life style of rather private but very privileged figures of militarily oriented nuclear, space, aircraft, and other fields of technology and science, which was established during Stalin's time, as we mentioned earlier, has remained almost unchanged to this day. As a rule, even the names of these experts are unknown to the general public. All the prizes they receive (Stalin [later Lenin] Prizes in science and technology), orders, titles (Hero of Socialist Labor), and other honors are given to them secretly. (For scientists and experts in non-secret fields, all these awards are announced in the general press and associated with publicity.) The degree of secrecy varies. Some men, like Kurchatov, were permitted to have a more open life and even made foreign trips. However, this was mostly when their main task was already over. Others have remained unknown until they died and their contributions could be acknowledged only in obituaries. The most prominent are elected academicians. However, in the Academy membership lists neither their position nor other details are indicated. They live in luxury apartments, and have large country houses which are usually under guard. When they want to spend some time in the "outside" world during vacations, they receive false identity documents and travel with permanent

bodyguards. Even when they go swimming, the bodyguard swims nearby. They are forbidden to make any contact with other people (visits to friends, etc.) without preliminary notice to the KGB's special department. These arrangements are intended as a precaution against possible security risks, but they also have some political meaning: it is much easier to keep an unknown expert obedient, to deprive him of possible authority and influence in the decision-making processes.

S. P. Korolev, who recognized clearly his own importance and contribution to the Soviet space program, often tried to rebel against the anonymity of his official title (he was named in the general press simply "chief designer"). Nevertheless, his name was never mentioned in the press nor was his picture published before his tragic death on the operating table, in 1966.

In any case, the whole situation was well enough accepted and even appreciated by others who worked in such conditions. This special treatment created for them some kind of artificial independent world, a state within a state, with its own values, priorities, and customs. They had all they needed and never had to think about expenses, taxes, care for their children, or many other problems of life which occupy scientists in the same field of work, in Britain, for example.

In some ways this isolation was a kind of extension of the scientific "prison institutes," but in a most luxurious form. The experts who were at the top of this system were not just the elite of the elite in science, they were part of the ruling elite as well. (This example was followed in China, where research workers in the nuclear and missile fields lived in an artificial, isolated world of their own.) This policy certainly paid off well, albeit at the expense of the population as a whole.

The policy of *détente* certainly could make the development of retarded branches of Soviet science, technology, and economy easier and less expensive, but it is wrong to think that *détente* will seriously add to Soviet military power and the danger of military aggression. In a highly centralized state, military development is virtually independent of the general standard of life or of international cooperation in other branches of science. Military technology is more often stimulated by bad relations than by cooperation. In a country like the USSR, emphasis on the increase of military power does not determine whether the workers are well- or ill-dressed and fed. A reduced standard of living can easily coexist with a large military budget and a luxurious life for the small minority.

A country whose ordinary people have more wealth, more economic stability, and more entertainment is usually more peaceful and politically safe for its neighbors. It is, of course, a more influential country as well. It is just this *influence,* rather than the military aggression, which is what the critics of *détente* probably do not want to see. I would like to make it clear, however, that the influence of the USSR will grow in either case—with *détente* or without it. *With* détente, *the influence is mutual and the situation is safer*. Without *détente,* the influence would still be felt, but through military means.

Critics would again insist that the USSR and its socialist regime as such are unchangeable, closed to external influence. I am not disputing this view by political or social arguments. I will try to talk about science only and will try to find out whether something has really changed in Soviet science during 1972–1976, a period when the policy of *détente* was slowly and with restraint applied as an experimental approach to the relations between East and West, or, more accurately, between the socialist and capitalist economies.

Trends in Soviet Science and Détente

Objective study of trends in Soviet science under *détente* is rather difficult, because assessment must be based on facts, features, and developments which are multiform, and are not clearly connected with political decisions. Many internal decisions are secret and can be felt only through their social consequences. For example, we may suppose that a specific decision was certainly made to control the emigration of Jewish scientists and intellectuals. This decision has made emigration easier for research workers in the humanities (history, psychology, linguistics, literary criticism, etc.), modernistic painters, writers, and other intellectuals who are not involved in the technologically "productive" areas and who encounter more difficulties in their settlement in the West and in Israel. The same is probably true for theoretical physicists, mathematicians, biologists, and medical experts, especially from troublesome dissident groups. But on the other hand, emigration for experts in technology and practically oriented research fields is still very difficult, even more so than it was in 1971. (I am not talking about emigration of research workers in "classified" research and military fields, as all decisions for such applicants are made on an individual basis.) However, the regulations covering these different categories of emigration are not only secret but are probably simply pragmatic, being decided periodically at a high level and transmitted to the local OVIR* offices by telephone from the KGB.

The "brain drain" is not viewed too seriously by the power structure of the USSR when Jewish scientists who are experts in history, literature, cinema, journalism, modern abstract art, photography, languages, psychology, law, medicine, and so on, emigrate to the West or to Israel. In emigration these experts often suffer frustration, disappointment, and homesickness; they also are handicapped by their ignorance of foreign languages and Western customs, and different professional standards in other countries create many difficulties in their employment (60). Many experts in medicine apply for emigration not knowing that Soviet medical diplomas are not valid in the United States, where they intend to go.

Mutual disappointment creates various advantages for Soviet

* Otdel Viz i Registratsii (Department of Visas and Registrations), a department within the regional or local republic divisions of the Ministry of the Interior usually attached to passport offices and responsible for arranging private trips abroad. Business trips abroad are arranged through the Foreign Relations Department of the different ministries or academies.

propaganda. As emigration is controlled, irreversible (through confiscation of passports), and selective, the state now manipulates emigration opportunities for its own convenience, often just to rid itself of dissidents, the old and the useless, the not too bright, and the unemployed. Many intellectual dissidents are disposed of by the stream of Jewish emigration. Such manipulation of emigration opportunities cannot be merely spontaneous—it constitutes the result of specific decisions, at the same time compromising with Western demands and conditions about freedom of emigration, clearly using the new possibility of reducing internal opposition and dissent, and sometimes avoiding an oversupply of labor in particular professions.

The problems of emigration and the "brain drain" are, however, outside the province of this work. They show that some trends which also marginally involve science have occurred during the last few years, and have become factors with indirect importance for other internal problems of science. Even given the clearly established trend of restricted and mostly Jewish emigration (varying between 10,000 and 35,000 people per year), it became difficult to avoid tension in the whole scientific community which has always existed over the problem of foreign travel for professional reasons (to attend congresses or conferences, or for temporary research work abroad). There are no valid figures available for the number of Soviet scientists who were able to travel abroad during 1975–1976 in comparison with 1969–1970, but at least in biochemistry I can be a witness to some relaxation. At the tenth meeting of the Federation of European Biochemical Societies in Paris in 1975, there were at least one hundred Soviet biochemists—a number unprecedented for any such meeting before. (Most of the Soviet participants were registered after arrival, but this is usual, because the selection of participants could not be finished before the deadline for reduced-fee registration in advance. The "luxury" of preliminary registration is available only for Soviet academicians and a few other celebrities.) At the Tenth International Biochemical Congress in Hamburg in 1976, the number of Soviet participants was smaller, but their pattern was very random; there were many young scientists from provincial institutes—a case which would be difficult to imagine in 1969–1970. I talked with some of my Soviet colleagues, and they explained that the smaller attendance at the Hamburg meeting was simply because of the higher attraction of Paris as a place to visit, as the Soviet government still treats such travel to attend scientific meetings as tourism, and most of the Soviet participants pay the full cost of the trip (which is usually higher than their monthly salary even for senior scientists).

Of course, there are many more scientists who would participate in the international science meetings if arrangements for them were easier—and to some extent they have eased. However, the bureaucratic arrangements, application processes, and so on, are still the same. Because substantial emigration has become a well-known fact of current social life, restrictions on foreign travel have inevitably had to be relaxed to ease the frustration of refusals which often led to emigration demands. It is also necessary to point out that scientists who are placed in the "travel permitted" group become reluctant to join or support the dissidents.

If figures on scientific travel permits are not easily accessible, there are some other objective signs of a slow integration of Soviet science with world science. As a rule, I feel more competent to discuss the situation in biology, preferably in biochemistry and gerontology, but the changes in these fields are certainly not isolated from the general trend.

For example, the number of publications of Soviet authors in the international (mostly English-language) biochemical and biological journals has increased significantly during the last six or seven years. This trend can be observed just by looking through the author indexes of various journals. This increase is probably a spontaneous process, associated with many factors, including, for example, the much higher standard of knowledge of the English language among Soviet scientists resulting from improved foreign-language education in the USSR. However, the higher quality of research work may be responsible as well, because it makes contributors more confident about the possible acceptance of their papers by foreign journals. Another reason may be the deliberate actions of research workers themselves who would like to have their work published for the international scientific audience as quickly as possible. One may also suggest that the frustrating internal procedure for submitting a paper for publication abroad has become easier than it was six years ago. But this I doubt very much and prefer to think that more and more authors simply ignore these procedures since the direct publication of research work abroad is no longer such a big political and professional risk as it was ten or twenty years ago.

The delay in the publication of works submitted to Soviet scientific journals was a cause of severe criticism for many years. The demand for the creation of journals to provide speedy publication within the Soviet Union has not yet been met, nor has the demand to create international journals within the USSR which would publish papers in English. (Most scientific and scholarly journals in Poland, Romania, and Hungary now appear in English.) A quick selective comparison for the end of 1976 provides the following picture of comparative speed of publication in the USSR, in Europe, and in the United States.

Average Time Elapsed between Receiving Papers and Their Publication in Comparable Soviet and Foreign Journals in July–September, 1976

USSR	Europe and USA
Biokhimiya, 10–11 months	*European Journal of Biochemistry*, 4–5 months
Molekularnaya Biologiya, 14–16 months	*Molecular Biology*, 6–7 months
Genetika, 11–12 months	*Genetics*, 6–7 months
Zhurnal Evolutsionnoi Biokhimii i Fiziologii, 18–20 months	*Journal of Comparative Biochemistry and Physiology*, 5–6 months
Ontogenesis, 12–14 months	*Journal of Developmental Biology*, 5–6 months

This list could be continued indefinitely and the picture would still be the same. In 1967 the lag for Soviet technical journals was even higher (39).

In the USSR, as I have already indicated, there are no biochemical or biological journals specializing in *rapid* publication in biology. At the same time, in Europe and the United States there are several such journals, and high-quality papers in biochemistry can be published in *FEBS-Letters* within two to three months, and in *Biochemical Biophysical Research Communication* within two months or a few weeks. These advantages of rapid publication have made Soviet biochemists turn their attention to the West, where they would like to make their work known even to their Soviet colleagues as soon as possible. The increase in Soviet publications is most evident in the *international* journals, like *FEBS-Letters* or *European Journal of Biochemistry,* but not in the national English or American journals like *Biochemical Journal* (Britain) or *Journal of Biological Chemistry* (United States), which also do not discriminate between contributors from different countries. I think that this is related to the obvious fact (and another significant change during the last five or six years) that many international journals which publish papers in English now have Soviet scientists among the members of their editorial boards or even as regional editors (every such journal usually has a managing editor and several regional editors). The participation of Soviet scientists in editorial work in foreign journals is a matter for approval by the academies of sciences, ministries, or other high authorities, which cooperate with the Party system and censorship as well in making their decision. But the personal invitation to participate arrives from abroad, from the managing editor or editor in chief (or sometimes from the publisher). Therefore such offers are usually received by scientists who are well known in one or another professional field. The widespread notion that such permission from the Soviet authorities is given only to members of the Party is not true. From the list of a few names (given below), which I have selected at random from the editorial boards of some international biological journals, it is evident that more than half of the Russian scientists are not members of the Party. All are experts in their field and I personally know almost all of them. (I do not indicate their titles, but all of these scientists are either academicians, corresponding members of academies, or at least professors, but this can be expected anyway if we talk about prominent figures.) They are A. S. Spirin and Yu. A. Ovchinnikov (editors of *FEBS-Letters*); G. P. Georgiev (editor of *European Journal of Biochemistry*); N. P. Dubinin, S. E. Bresler, and N. W. Luchnik (members of the board of *Mutation Research*); O. E. Epifanova (member of the editorial board of *Experimental Cell Biology*); A. E. Braunstein (editor of *Analytical Biochemistry*); S. Kafiani (member of the editorial board of *Differentiation*), and so on. This list could be extended to several dozen journals published in the West with Soviet participation in their editorial boards (*Cytogenetics & Cell Genetics, Biochemical Pharmacology, Experimental Gerontology, Journal of Molecular Evolution, Gerontologia*, etc.).

In a recent article by émigré scientist Y. Levich, "Trying to Keep in Touch" (*Nature*, vol. 263 [1976], 366–367), the author describes the many difficulties faced by foreign contacts of Soviet scientists. Many statements in this article are too generalized; they could be typical for workers in a research establishment cleared for classified work. The author correctly describes the institute's control of papers which their authors want to send abroad. However, Levich is wrong when he writes: "A particular form of control has appeared with the inclusion of Soviet representatives on the editorial boards of some international magazines, providing an opportunity of preventing the publication of articles in foreign magazines." First, the "inclusion" of Soviet scientists on the editorial boards of international journals is not a Soviet decision, but the result of invitations from the foreign managing editors who decide on and select the representative editorial board.

Soviet members, like other members of the editorial board, have no veto powers, and the acceptance of an article, whether submitted from the USSR or from another country, depends entirely upon the referee's or reviewer's recommendations. While working in the USSR, I myself was invited to join the editorial boards of two international journals, *Mechanisms of Aging and Development* and *Journal of Molecular Evolution,* and in both cases the invitations arrived from the editors of these journals. My task was not to veto but to encourage contributions from the USSR. No one among the scientists whose names I mentioned as members of editorial boards is able to exercise censorship in any form. If they were given a paper to consider for publication abroad, their decision would be based on merit and quality and they would certainly separate the good from the bad.

It is very helpful to criticize many aspects of Soviet science, but it is counterproductive to falsify reality and to exaggerate existing difficulties.

In spite of better opportunities today to publish research abroad, the majority of Soviet scientists still publish their papers in the Russian language. (Ukrainian and other languages account for no more than 10 percent of scientific publications. However, such publications have, as a rule, local significance only. Hardly anybody in the Moscow research institutes can read technical-scientific literature in Ukrainian, Armenian, or Latvian.) The total number of scientific and technical journals in Russian (according to the catalogue for open subscription *Gazety i Zhurnaly USSR*) was about 550 in 1975. There are several dozen so-called "closed" journals, which are not listed in the open catalogue and where results obtained in the classified fields of nuclear physics, missile research, and so on, are "published." Such journals have a very limited circulation, and each copy is numbered (often the circulation is between 200 and 300). To read such journals, one must be an expert in the appropriate classified field of research, in order to possess clearance of the highest category and to have access to such journals in the reading rooms of the "special departments" (*osoby otdel*). Through these classified periodical publications, research workers in secret fields of research can exchange information, and can now obtain the moral satisfaction of getting their work published. This classified printed journal system has replaced the similarly closed individual report system which dominated the secret fields of science and technology between 1946 and 1966, and has in some measure alleviated the dissatisfaction among both junior and senior personnel by providing some facility for publication.

The number of Soviet authors who have published their books abroad has similarly increased during the last five or six years. In many cases, this is related to arrangements made through official channels, but quite a few social scientists have published their work in the foreign press—even in Russian, but abroad, without any censorship and through direct, often confidential communication with the publishers. Russian-language books published abroad have increased many times between 1970 and 1976, and only a few of them are by authors who have emigrated abroad. Even former and dismissed Communist political leaders (Khrushchev, Dubček, Gomulka) have published their works and memoirs abroad, taking advantage of the opportunity to escape censorship.

All these new trends were certainly encouraged by the spirit of *détente,* and many of them are spontaneous and not officially programmed. Soviet participation in the Universal Copyright Convention (since 1973) was the decision of the Soviet leadership, but the special organization created to deal with new problems in the exchange of information through publications and translations is not able to control all aspects of the situation.

In technological areas, the *détente* policy has certainly accelerated the use of foreign technology and equipment both for research and for industry. In the USSR, the tendency to develop domestically every type of equipment which exists in the world has started to decline. The statistics (58) clearly indicate this trend during the last five to six years. The Soviet government today considers it more economical to acquire the license to manufacture many types of foreign equipment than to try to "invent" something comparable within the USSR. This solution is not only more economical but most probably provides better-quality equipment for research and technical use. It strengthens and modernizes Soviet industry, but at the same time makes it more dependent on foreign technological exchange and spare parts. This dependence, however, is mutual and, given more developed cooperation in the future, the Soviet market for foreign technology may be able to stabilize the producing country's economy (through cooperation) or to destabilize it (through non-cooperation in the case of some political crisis). As could have been predicted, real cooperation in science with the West had to start from the field of space research, which probably was incomparably more expensive and resource-consuming than all other fields taken together. The USSR, as well as the United States, was no longer able to afford rivalry in that area. Both countries had cause for satisfaction: one with the first satellite and the first manned orbital space flight and the other with the first men on the moon's surface. The spectacular effects of both space programs were certainly enormous and the psychological impact very strong, but the practically useful information per million spent was most probably the lowest among all research fields in modern science. Nearly twenty years of competition had been wasteful enough.

Space research programs certainly gave *some* positive scientific information (the study of the solar system, high-energy rays, astrophysics, meteorology, and some others). However, even the Americans have acknowledged (61) that scientific research in itself "never has been the principal justification for space expenditures. Two main factors which were on the USA's side in support of space programs were: 1) the popular appeal of space adventure, and 2) the enhancement of national esteem and international prestige," which were typical for the Soviet space program as well. *Détente* is a good excuse for the end of rivalry in this field and an opportunity for cooperation. It also is a good excuse for contraction of the whole program and a decrease in the space budget, which both countries have very much needed. Among the many advantages of space research cooperation in both countries, the reduction of mutual expenses in this too costly field is probably the most advantageous. The attempt of the United States to restrict cooperation and preserve a unilateral lead in technology required for advanced space experiments is, I think, a temporary phenomenon. With the end of the competitive mood, the fiscal constraints and rate of decrease in the budget for space are much greater in the United States than in the USSR, where government expenditures are not under any real control from the Supreme Soviet. (American space spending in fact has declined rapidly since 1969. Soviet space spending, according to U.S. expert estimates [61], has not declined and remains stable to date.)

If the "duplication" trend during Khrushchev's time made it necessary to create an efficient scientific information service, the new "integration" stage has made this service even more important. For example, in chemistry, twenty years ago, in order to introduce a new chemical for practical use, it was enough to synthesize it according to a method developed elsewhere. The chemical would be used within the USSR, and nonparticipation in the patent and license convention made such "pirate" development easier than the attempt to find another chemical with identical or closely similar properties. With adherence to the convention, the only three options that were open were either to buy the final chemical, to buy a technique for its synthesis and production, or to develop a new chemical with the same properties of use (potentially there are often very many substances with the same capacity to absorb, dissolve, filter, separate, explode, etc.). In this case, it would be necessary for the chemist not to repeat a process already somewhere in use, but to ensure that his new ideas were not already explored and licensed somewhere else.

Anyway, every year about 100,000 new chemicals are described and about 4 million are already in store for potential use. There are, for example, more than 100,000 substances for fluorine only. Such pressure of *existing* knowledge has made it necessary to develop informational facilities according to modern standards and to use widely complex computer systems. This happened to be the most backward field of Soviet technology—a posthumous effect of the ideological condemnation of cybernetics as a bourgeois science. Computer information centers have become usual for many large research institutes and industrial complexes. A comprehensive computer system was created in the All-Union Institute of Scientific and Technical Information and in many other information centers. However, Soviet-made modern computers are grossly inferior to the new models produced in the West, and the whole computer industry is still backward. Among various types of equipment, computers and electronics in general have now become the main Soviet import demand and probably will continue to be so for at least a decade.

In order to reduce this gap, the educational system has made electronic and computer engineering the priority branch in higher technical education, and it could happen that the gap will be filled earlier than I have predicted. The gap in atomic weapons was closed much more quickly, but the technical problem of making a bomb had a ceiling which does not exist in making modern computers. Also, nuclear physics in the USSR was in very good shape when the country started the nuclear race. Cybernetics and computer technology simply did not exist when it became necessary in 1955 to follow well-advanced Western electronics.

This is one of the reasons why, in spite of many Russian-language comprehensive abstract journals in chemistry, biology, physics, and other sciences, and their very modest prices, which easily permit individual subscription, those who know English usually prefer to use American abstract information if it is available (*Chemical Abstracts, Biological Abstracts,* etc.). American abstract journals have good systems of computer-compiled author and subject index supplements which made the search for necessary information much easier. Russian abstract journals from the very beginning of their existence felt the need and have promised many times to bring out such indexes, at least as annual supplementary volumes, but this has not happened yet. As all indexes for books and journals are still handmade by human labor, the Soviet versions of *Chemical Abstracts* and other journals of this kind are not able to make indexes in time. Moreover, when they do become available, with a five- or six-year delay as a rule, they are too old and have very limited use.

The internationalization of Soviet science and technology, which is both a cause and effect of *détente,* is still in the very first stages. The integration gives the large scientific community in the USSR closer connections with the West and Western values and standards of living. The "Westernization" of the Soviet intellectual community is a visible process, but the increase of tolerance toward Communist countries had a certain influence on the development of the left and socialist trends in the West as well.

I do not plan to discuss how the West regards the new situation created by increasing cooperation with the USSR and other countries of the socialist bloc, but it is necessary to see how Soviet leadership reacts to the internationalization of Soviet science.

New Forms of Political Control

Academic positions in the USSR during the last few years, as we have seen earlier, were more and more associated with many privileges: a higher standard of living, better access to foreign literature, officially permitted communication with foreign colleagues, exchange of reprints, participation in international conferences and meetings within the USSR, and, for the lucky ones, even trips abroad. Research scientists need special training and many years of extra education. This they need comparatively high financial support from the state, involving in many cases foreign currency. Each research scientist costs the state and the "taxpayer" * a great deal of money.

It became popular to proclaim that science is now also a productive force like industry. However, many theoretical fields do not produce immediately visible returns, and, of course, much research returns nothing at all, either because of repetition or because of research in the wrong direction.

Research work requires honesty, integrity, a high level of intellectual and educational standards, and a lot more. Even in the totalitarian states research work needs comparative freedom of choice for scientists and cannot be treated in the same way as other fields of human activities. But where individual freedom and human rights are not highly respected, and where state interests are absolutely dominant, political leaders want obedient scientists and loyal research workers who do not interfere in political life. The USSR wants not just obedient scientists, it wants trusted and actively loyal scientists. The research elite has to be an integral part of the political and governmental elite and to share all values of socialist ideology in the political area. The state wants scientists to be *patriotic* not only from a national point of view but also from a political point of view, devoted to the idea of building a prosperous socialist state. Scientists in the USSR are expected to be loyal not only to their respective sciences, not only to the national values and traditions of their Motherland and people, but to the existing political system as well.

Unfortunately for political leaders, research work is bound up with the challenge of dogmas and authorities, disputes, confrontations of conflicting views,the testing of ideas and theories, and comparative freedom of thought. A good scientist cannot just believe in something. These special conditions and qualities produce a higher frequency of political dissent in science than in all

---

* "Taxpayer" in the USSR is a relative term. Direct taxation from salary and income is usually low. But indirect taxation, through reduced salaries and high prices for many commodities and goods, is substantial.

other fields of social life. *Détente* and the relaxation of repressions inevitably increase this frequency, and though this possibility was not foreseen some time ago, it certainly became the case as soon as the *détente* policy was implemented on even a modest scale. In the USSR social forecasts in most cases were incorrect, and leaders usually tried empirically to control in one way or another the current situation and processes which were already evident and well developed. The same happened with political dissidence in science—the leaders faced the problem when it became an international issue.

I do not plan to discuss in detail here the methods and actions which have been used and are still being used to suppress political dissent in science. They are approximately the same as those which were used before the *détente* era, but with a change of pattern. The "soft" methods like forced or encouraged emigration, demotion, "warnings," and others are more widely used than before. "Hard" methods are now used less often. What is more important are the measures designed especially to *prevent* dissidence in science, to recognize the potential dissidents and to prevent them from attaining prominence or even from following a scientific career. These preventive measures would certainly mean the loss of some talent, but with a highly redundant scientific community and the decreased role of individuals in many fields of complex research, such a loss is not considered serious. The methods of political and ideological control in science were not introduced by a single and carefully evaluated decision. They were more a result of empirical tests of different approaches, old and new, and their future effect is too difficult to predict at the moment. I am personally rather skeptical about their real effectiveness, but I shall consider and explain them in any case.

The main purpose of some new regulations introduced in 1972–1976 was to prevent potential or actual political dissidents from obtaining prominence or a high rank in science or in any other field of intellectual activity. From the general point of view, these regulations could be divided into two main categories, unofficial and official. *Unofficial* measures take many forms and affect individuals. It is not always easy to prove that they in fact exist, and any Party or government figure would certainly deny their very existence. *Official* measures exist in the form of obligatory instructions and regulations that must be followed. They could be distributed in the form of special instructions for administrators, but could also be published and discussed in newspapers and journals as rules regulating professional relations in research and educational systems. As official regulations in some cases are just legalized unofficial rules which existed earlier, I will try to describe briefly some of these unofficial rules. When I classify them as "unofficial," this means that there are no *open* written rules to apply, and the given instruction and decision is a result of a personal recommendation, which, for example, the regional Party secretary could give to the director of the research institute or the rector of the university or the chairman of the university Party Committee about a certain person. The advice could be given through the KGB system as well, because in many research and educational institutes there are "special" or "first" departments which have dual connections. One of them is with the KGB.

In Soviet higher education there was almost always some system of *discrimination* against certain groups of applicants, who would find it more difficult (but usually not absolutely impossible) to be accepted as undergraduates of the university or one or another of the higher schools. To make all kinds of discrimination easier, each application was usually accompanied by very long and very thorough questionnaires (*ankety*) and by an autobiographical essay. (Application questionnaires are normal everywhere, but hardly anybody in Britain or the U.S. would expect questions about arrested relatives, relatives abroad, residence in "territories occupied during the war," etc.). During the short post-Revolution period, "class origin," presenting difficulties for applicants from bourgeois classes, children of "white" officers, or religious men and women, was a most important pretext for discrimination. Representatives of the working classes and poor peasantry were given preference in higher education.

During Stalin's time discrimination was much wider, and the children of the more wealthy peasants (kulaks), who had been deported during collectivization, and the children of millions of people arrested during various waves of terror had much more difficulty in finding their way into higher education than others. After the war discrimination was introduced against those who had lived at some time under German occupation, as well as against the Jews, the Volga Germans, and some other national minorities (Tartars, Kalmyks, Chechens, etc.). At the same time, there were many exceptions, and the road to higher education was not completely blocked for the young generation of any particular group. The prestigious universities (like Moscow or Leningrad), or special higher schools (like the Higher Institute of International Relations, aircraft institutes [colleges], physics-technical faculties, etc.), were closed for certain categories of applicants. However, it was impossible to discriminate against those who lived in German-occupied areas when the universities themselves were situated in the cities which had been under German occupation for a long time (Rostov-on-Don, Kiev, Minsk, Lvov, Odessa, and many others). It was difficult to discriminate against young demobilized soldiers or officers after the war (even those with a "bad" class origin, or whose parents had been arrested), because the number of male applicants for higher education was small, and in many universities and higher schools, out of ten or twenty applicants only one would be a man. The war had seriously depleted the generation of young men, and a talented boy or demobilized soldier could find a possibility for higher education and a scientific career in spite of any official or "unofficial" rules.

These forms of "class" and social discrimination became irrelevant in 1971–1976. But some unofficial (though not total) discrimination—for example, against the Jews (as potential emigrants)—did exist. The discriminatory approach became more individual and less obvious. To give some valid reasons for "proper" selection independent of the results of examination competition, the new system had been introduced as valid and equal with examinations. (Entrance examinations are compulsory for higher education in the USSR, and all examinations, either written or oral, are signed, not coded. Therefore the examiners know the names of the authors of the essays.)

Two additional "tests" were the interview and the *kharacteristika* (character reference) from secondary school. (Secondary school in the USSR is equivalent to high school in the United States.) In the *kharacteristika* from school not only the academic but the ideological, moral, and political qualities of the school leaver had to be indicated. Membership in the Komsomol (Young Communist League) was also important, but not obligatory or decisive. The official constitutional doctrine that the Party is a "leading group" of the working class and the people, and the Komsomol a "leading group" of the young generation, made it unnecessary that *all* workers, staff personnel, or students be "card" members of either the Party or Komsomol. If these organizations were considered "leading," they had to lead others, not assimilate them all. This principle of "leadership" of the Party is usual for any social group, and the widely held notion that it is obligatory for heads of laboratories, directors of institutes and academies, etc., in the USSR to be members of the Communist Party is wrong. There are quite a few academy institute directors and academicians and heads of departments and laboratories who are not members of the Party and who are not advised to be, if they apply for membership. The same is true for even more politically oriented groups, such as elected delegates ("deputates") of Supreme Soviets and local Soviets, lawyers, editors, journalists, and many others. Each social group must consist of members and non-members of the Party, leaders and those who are led, supporters of the Party but representing the non-Party segment of the population. The "union" of members of the Party and non-Party masses is customary for any social or professional group, except on the government level.

This "pre-educational" selection stage is not yet really important for political control of dissent in science. The political views most often develop later, either during higher education or during research work when the person is more mature. In the Soviet Union, education at all levels includes political education as well: general study of society in secondary school, Marxism (philosophy and political economy) and the history of the CPSU at the higher educational level. Alternative political education is simply not available, and literature with modern non-Marxist political discussions is not published for sale. This is one of the major factors in the very low incidence of political dissent among students and undergraduates. The critical re-evaluation of political dogmas starts later on when the basic education is over and a young boy or girl starts on his own in life (or in science).

It has been said many times recently by such well-known dissident figures as A. Solzhenitsyn, V. Maximov, A. Sakharov, V. Bukovsky, and others, especially those who have now emigrated, that Marxism is not believed in any more in the USSR, and that even young students in secondary schools do not believe it and "laugh about it." "In the USSR today Marxism is considered so low that it has become an anecdote; all hold it in contempt. Nobody, not even students and schoolboys, talks about it seriously, without a smile" (Solzhenitsyn, 62). This, however, is simply not true. Some intellectuals may become disillusioned with "classical" Marxism and may start to look for alternative ideologies, sometimes even religious, but such trends are not common. The major part of the Soviet intellectual community still frankly remains within the framework of the Marxist ideology, but it tries to adapt the Marxist analysis of modern society to the changing realities and to favor more liberal, reformist ideas of socialism. Marxism as methodology and socialism as the structure of social life are not the political ideas which Soviet scientists believe because of official education only. These ideas have a logic and scientific appeal which could attract scientists—and not in the USSR alone. Religion in its various forms does not have the same appeal for the majority of Soviet intellectuals. The "free" countries, like the United States, France, and some others, do not have any ideologies comparable with Marxism, but they have many other values in the form of a special respect for basic human rights, democracy, and civil liberty. The endeavor to link these values of the Western world with the values of socialist models of the "just society" is the main line which the mature political thinking of many Soviet scientists takes. In this direction they probably approach social-democratic principles. Social-democratic parties and governments in the West in the past had programs with their background and roots in Marxism. The Soviet "liberal Marxists" (or "democratic Marxists" or "neo-Marxists") probably never reach "social-democratic" standards, because too many changes in the USSR have already become irreversible. But at least many of them do not consider the current official Soviet system as a true and just socialist system. They would describe it as a transitional phase developing from the dictatorial (totalitarian) pseudo-socialism of Stalin's type, toward a more democratic and just society with a real respect for human rights and universal moral values. The official Party view is different. It considers that this transition from a totalitarian Stalinist regime toward real socialism has already been completed. The more common and yet dissident view is that this transition has been cautiously started and that there is still a long way to go. On the surface of Soviet science, the official view would seem to be dominant, but under the surface debates are common and often intense. The government and the Party leadership are fairly tolerant of these mostly private debates provided they are not expressed openly in publications, either at conferences and meetings or more especially in the foreign press.

During Stalin's time, obedience and conformity were obligatory in both official and personal circles, private groups, and even in family life. Now personal nonconformist views can be expressed without reservations in private groups, but support for the *general* line is obligatory in official circles. Outspoken dissidence is not tolerated, and because the scientific community is the most likely source of such dissidence, a complex of different measures has been (mostly empirically) developed over several years to keep the majority of scientists within the official conformist bounds of political behavior. I will not discuss here the "case histories" of the different regulations, but simply explain some of them and try to show how they work.

POLITICAL CONFORMISM AND THE SYSTEM OF ACADEMIC QUALIFICATION

In the USSR a successful research career depends upon academic degrees and titles (candidate of science, doctor of science, professor, and some others) much more than in many Western countries with their multiform character of scientific and research systems and the comparative independence of individual universities and research units. All scientific units in the USSR are organized in a hierarchical administrative pyramid system based on the kind of research or the educational qualifications, degrees, and diplomas necessary (or often obligatory) for any particular position. The distribution of opportunities is too complex to discuss here in detail, because it depends on too many factors; for example, the category within the pyramid (Academy of Sciences of the USSR, Academy of Medical Sciences of the USSR, union-republic academies, ministerial research network, provincial research units, universities and other higher schools, etc.). To be the head of a laboratory or department within the central Academy system or to be a professor of the university, the doctor of science degree is usually necessary. However, within the union-republic provincial academies or within the research institutes of specialized academies (medical, agricultural), either the doctor or candidate degree is acceptable for appointment to the position of head of a laboratory, simply because there are not enough doctors available for all possible scientific-administration positions. At the lowest category of the research system—in the network of provincial experimental stations (agriculture, plant breeding, soil research, etc.) or at the laboratories directly attached to the industrial complexes, clinics, and hospitals (*zavodskaya laboratoria,* industrial laboratory; *diagnosticheskaya laboratoriia,* diagnostic laboratory; etc.)— even applicants with candidate degrees are not always available, and many heads and chiefs of such laboratories hold a professional diploma only (engineering, agronomy, medical practitioner, and others). In this case, however, the salary of the head of a large industrial laboratory could be only a quarter of the amount which the head of a small *academic* laboratory (with a doctor of science degree) is entitled to receive. The dependence of income on titles and degrees makes it inevitable that tens of thousands of research workers unceasingly strive to obtain research degrees through examinations and the preparation of theses which could be considered appropriate for candidate or doctoral degrees. The Soviet system of research qualification opens such possibilities for all graduates with a professional diploma, and the thesis can be prepared not only during postgraduate work (*aspirantura*), but at any research or research assistant position, without leaving one's main job.

The procedure for awarding academic degrees has already been considered. Although it is complex and looks even more formidable, multistaged, and bureaucratic in comparison with Western standards, the complexity of the system is presumably designed to ensure the high quality of scientific research. However, this multistage system is more and more used not only to assess scientific merit but for political ends as well.

I do not plan to discuss the history of the political misuse of the system, which during Stalin's and a part of Khrushchev's time as well was often a field of struggle between sciences and pseudosciences. Even the highest-quality work in classical genetics in all fields of biology (biochemistry, botany, zoology, plant breeding, etc.) did not stand a chance of passing through. It was blocked either at the stage of "open defense" or (if the applicant who was free to select the university or the research institute for consideration of his thesis found a favorable place for "defense" and got the preliminary decision about the degree by secret ballot) at the stage of confirmation of his degree by the All-Union Qualification Commission at the Ministry of Higher Education. At that time, for example, the Novosibirsk University was a good place to defend a thesis in the area of classical genetics, as many geneticists purged from Moscow moved east. However, the commissions of experts and the plenary commissions of the central Moscow establishment at the ministry had T. D. Lysenko himself as a prominent member, and many of his followers were members of all the groups of experts. Work that successfully qualified for a doctor's degree in biology at universities could be rejected at this level, and the final awarding of a doctor's diploma denied. I could describe dozens of such cases before 1965; since 1965, when Lysenko and many of his followers were removed from the Highest Qualification Commission, although the rejection of a well-deserved qualification in genetics has been less common, there is still the possibility of discrimination for other reasons. As the same commission considers and confirms titles as well (senior research scientist, assistant professor, professor, etc.), its control over the life of the whole scientific community must be evident.

However, this system of control was not good enough to prevent dissent in science, and during the last six or seven years several major reforms have been introduced to increase state and political control over scientific qualification awards. The rules (*polozhenie*) of scientific qualification have been reconsidered several times. Before 1970 the Highest Qualification Commission and its groups of experts seriously considered only the top titles and degrees (doctor of science, senior research scientist, and professor). The degree of candidate and the title "Junior Research Worker" were under control, but consideration was by sampling. A few probably not randomly selected works and applications were picked up for analysis, but all the others were automatically confirmed, because the commission simply was not able to analyze seriously so many works (10,000 to 12,000 theses and about 30,000 qualification titles every year) at this level.

Since 1971, when the commission was reorganized and enlarged, it became obligatory to reconsider all degrees and titles, independently of their "level," and confirmation of the most common candidate degree stopped being routine. Instead of the four months previously allocated for such consideration (this four-month "maximum" was often violated in controversial cases), a ten-month period of "expert consideration" became officially possible for doctoral degrees. The new rules gave the commission the power not only to award or reject recommendations from universities or research institutes, but also to deprive scientists of degrees and titles which had been awarded in previous years *"in case of anti-patriotic and anti-moral behavior."* The power to deprive people of degrees in science existed earlier as well, but it was prescribed for use in such cases as plagiarism, falsification of research data, the use of "ghost" authors, and so on.

The meaning of the new terms "anti-patriotic" and "anti-moral" was broad, and at this time they were clearly directed against dissenters and Jewish scientists who had applied for permission to emigrate. However, it became evident very soon that the implementation of these new powers was not easy. The decision in each case was to be made after special hearings at the general academic council of the university or research institute and at the Highest Qualification Commission level and was then to be published in a special *Bulletin of the Ministry of Higher Education.* In all cases the academic councils of research institutes or "expert commissions," who would readily agree to discuss accusations of plagiarism or faked research, were reluctant to discuss the "anti-patriotic" or "anti-moral" behavior of well-established scientists, and the publicity given to such a decision would often have a reverse effect. Professor Efim Etkind's case, described in his own work (63), is a good example of the difficulties of depriving a distinguished scholar of the title of professor because of "anti-patriotic" behavior. Professor Etkind is a leading expert in Russian literature, and he was a good friend of Solzhenitsyn. Solzhenitsyn, while in the USSR, often called upon Etkind for advice and assistance. However, after Solzhenitsyn's forced deportation, this assistance was considered "anti-patriotic" (the information about his assistance came from the KGB, it was not common knowledge). Etkind's colleagues at the Herzen Pedagogical University, where he was one of the most popular teachers of Russian and Soviet literature, did not know of his friendship with Solzhenitsyn. Etkind's behavior as professor at the university in Leningrad was beyond reproach. His academic work was highly valued, and it was a great surprise and shock to his academic colleagues when the KGB prepared a "case" against him with a demand to expel him from the university and to strip him of his title and degrees as well. Under strong pressure from the KGB and Party officials, the academic council finally complied, but the university and the officials were disgraced more than Etkind himself. Etkind lost his job and titles, but he could not be deprived of his knowledge and qualifications. His only recourse, however, was to emigrate, and he is now (since 1975) a professor at the University of Paris, where he is again teaching Russian, French, and German literature.

There were not many cases of this type, and the authorities did not try to use the rules against those who applied for emigration (though the attempt to emigrate is clearly "anti-patriotic" behavior). The initial move in 1973 to strip the academician Andrei Sakharov of his title met with strong resistance in Academy circles. The Academy had not earlier considered the suggestion made by some physicists to deprive T. D. Lysenko of his title, in spite of the fact that the interests of science were not at stake in Lysenko's case and that a move of this kind would not damage the Academy's international reputation.

To reverse a past decision proved to be difficult, but *preventing* dissident research workers from obtaining higher academic degrees or scholarly titles would be simple. Two true stories— among dozens of similar cases—will show how the system worked.

K., who was already a Candidate of Physics (a degree similar to an American Ph.D.), successfully defended his new research work for a Doctor of Physics degree. The university where the thesis was defended and received support was not aware of K.'s dissident activity, nor was the provincial research institute where the research work had been done. This is why K.'s thesis for a doctor's degree was accompanied by an excellent *kharacteristika* (character reference) signed by the director and the Party secretary, where the "moral" and "ideological" qualities of K. were indicated to be correct and sound. K., however, was close to some Moscow dissident circles, and this did not go unnoticed by the KGB. It could be embarrassing to put pressure on a large university academic council, and this is why at this stage everything seemed to be going well. The university's decision was, however, not final. The actual degree and diploma depend on the decision of the Highest Qualification Commission in Moscow, whose deliberations—unlike "defense" at the university—are not open to the public. It is now clear that the commission , which is "All-Union," has access to confidential information following routine inquiries by the "special department" of the KGB relating to recommendations on each case, especially when a doctoral degree is in question. The expert groups of the commission are composed of prominent scientists, and they consider the works included in the lists prepared by the staff members of the commission, who are also partly responsible for the selection of referees. Some especially trusted referees could be given tips on the work of unwelcome controversial figures, and because almost every research work is open to challenge, it is often not difficult to get a negative report. However, it happened to be a problem in K.'s case, because his work was of exceptionally high quality. In this circumstance, K.'s work was just put aside. When the one-year "consideration limit" was over, K. inquired about the reason for the long delay. The reply was typical—"a too busy schedule"—and there is no way of complaining about the All-Union Commission. After two years' delay in the consideration, and after many inquiries, often through influential friends, K. got the message. His reaction was probably neither typical nor brave, but at least it was understandable from the point of view of those who give their research promotion priority. K. stopped meeting his former dissident friends, lost interest in *samizdat* literature, and made an application to join the Communist Party. His "behavior" was now quite different and more careful. Four years after "open defense," he was finally able to obtain his Doctor of Physics diploma and become a member of the Party. But he also changed his place of work to escape the contempt of some of his old friends, who were at a lower level in the scientific scale of rank. Two years later he became more flexible, and did not go out of his way to avoid dissidents when he met them. He also renewed his interest in reading moderate *samizdat* literature, but now his new prominence in science and his new position as the head of a large department were not at stake. Moderate liberalism was more or less tolerated as soon as K. proved that he had got the message and was able to be obedient. *Obedience* is more valued now than *real* thoughts, as long as one's real thoughts are kept completely private.

The case of A. was more complex. A. was head of a large laboratory, a candidate of physics with a good scientific reputation. His dissidence was more or less open toward his colleagues and his administration, and he demonstrated his independence of thought clearly enough. He was sure that his moderate prominence made his position safe enough. However, he was dismissed from the institute as soon as a favorable opportunity arose (reorganization, the excuse which cannot be challenged either through the courts or through the administration). A. could not find a job for at least a year, and was finally employed as a *junior* research scientist, in spite of his age and his high academic standing. He was no longer able to maintain the standard of living he was used to, and he decided that this was too high a price for outspoken dissidence. A. concentrated all his efforts on his research work, which was now in applied physics, and he was able to develop some new technological designs which were well received in related industrial fields. He did not, however, sever his friendly relations with some of his former dissident friends, and he retained his lively interest in *samizdat* literature. But this part of his life was well hidden, and his new colleagues in his new institute knew him as a devoted nonpolitical scientist. When the position of senior research scientist in the department where he worked became vacant for open competition, A. was recommended by a "selection committee" and by the administration, and he won the position as well as the title, both by secret ballot of the academic council. However, both the position and the title are subject to confirmation, first by the research division of the ministry and then by the All-Union Highest Qualification Committee. To the shocked surprise of A.'s colleagues, and to his own great disappointment, neither decision of the institute's council was confirmed. A. is still in a junior grade, in spite of his age, qualifications, and success.

In both these cases, as well as in many others, the officials of the commission or ministry faced the delicate problem of reversing the decision already made within scientific circles; such a necessity is often painful and embarrassing. The results of this policy are not very predictable. In many cases they could lead to obedience, but in quite a few cases to more outspoken dissidence and disillusion. It became clear that the situation must be changed so that politically unreliable figures could be frustrated at the very beginning of their academic or scholarly recognition. Officials would certainly prefer not to reverse but to *prevent* such recognition. This obvious trend was realized in the new directives about scientific degrees and titles which were introduced at the beginning of 1975 by the decision of the Party Central Committee and Council of Ministers of the USSR: *About the Measures for Further Improvement of Qualification and Attestation of Scientific and Scientific-Educational Personnel.* The decision as such was, of course, full of praise for the quality of research, scientific merit, higher standards of education, and no decision of this kind could ever be assessed at face value. It is better to judge the real targets and intentions of this decree according to its factual implementations, which have been partly discussed in the article by K. Gusev, the deputy chairman of the Highest Qualification Commission, which was published one year later (64).

As my main intention here is to analyze the political aspects of such developments, I will not try to predict the importance of the new regulations on purely scientific matters. A few parts of the new regulations must, however, be considered, in the context of the current discussion. The new decree made the Highest Commission, which decided upon scientific titles and degrees, independent of the Ministry of Higher Education, but subject to the instructions of the Council of the Ministers of the USSR. This structural change gave the commission higher stature but made it directly dependent on the government and Party Central Committee (as well as the KGB). The commission is now far more bureaucratic, more comprehensive, and clearly a political state body not covered by the scientific-educational character of the Ministry of Higher Education. The Party and the government directives have been further realized in the new *Regulations about the System of the Awarding of Research Degrees and Scientific Titles,* which was introduced by the decision of the Council of Ministers of the USSR on December 29, 1975. In these "regulations," special care was taken to ensure that research degrees and scientific titles could only be awarded in cases where high scientific standards were "combined with the good mastering of the Marxist-Leninist theory, wide cultural level, and active participation in political life, and following in all actions the principles of Communist morality."

The previous possibility that, in special circumstances, some scientists could be considered for academic degrees without special "defense proceedings" (because of their outstanding contributions or by a summary of their publications) was now dismissed as invalid. This old rule made it possible in 1966 to award degrees and titles to a group of distinguished geneticists, who during the thirty years of Lysenko's rule had had no chance of obtaining scientific qualifications, but nevertheless continued to work productively, though in junior research positions.

The most radical change, however, was in the obvious elimination of the secret ballot system during the decision-making process. The academic council, which earlier decided on titles and degrees by secret ballot, must now make two separate decisions. The first decision is adopting the *project of resolution* about the merit and importance of the work. This project of resolution must be made by *open* ballot. The second stage is the final decision about awarding the actual degree, for which a secret ballot is still required. As a difference in the results of the open and secret ballots could embarrass the academic council, a negative initial resolution would certainly lead to a negative decision about a degree as well. The new system of "defense" makes the opinions of the politically more influential members of academic councils—directors, Party chairmen, trade-union chairmen, heads of so-called "special" departments (the last three members of the council are voting members even when they are not members of the research staff)—much more decisive at the first stage of open voting on the general resolution. These new rules make it possible to eliminate the research work of dissident scientists, either at the preliminary stage (for not fully mastering Marxist-Leninist theories and ignoring the principles of Communist morality) or in the vote during the defense proceeding itself. The last possibility can be exercised when dissident activity is not open and is known only to the "special department" or the Party chairman but not to the rank-and-file members of academic councils. In this way the Highest Commission is able to reduce the frequency of the need to reverse the council's decisions.

The full text of the new regulations has been published (65), and I do not plan to comment on it here in detail. It probably improved the scientific standards of research as well; the government of the USSR is not indifferent to the quality and prestige of Soviet science. However, its political intention is clear enough: to transfer the responsibility of preventing dissidence among scientists from political leadership to senior authorities in the scientific communities themselves. There is the unchanged rule to the effect that if the academic council of the university or research institute awards scientific degrees to inappropriate candidates three times in the course of one year (i.e., awards which are not subsequently confirmed by the Highest Commission), such "errors" cost these councils their very right to consider and award such degrees at all for a minimum of three years. Clearly, there is vigilance at the gates of paradise—it is not easy to become a member of the scientific elite. Nevertheless, science in the USSR is still the main source of dissent and will remain so in spite of all these measures, just because a critical mind is essential for all experimentation and research.

POLITICAL CONTROL OF PROMOTION IN SCIENCE AND EDUCATION

Political control over the different paths of promotion and recognition of distinction in Soviet science has existed since the origin of the Soviet Union. However, its pattern, character, methods, and real influence have varied during different periods of Soviet history. For at least two decades after the Revolution, political pressure over many research units was largely *external*, because Party organizations within these units were still too weak to control efficiently the activities of their research centers. Small

Party groups (*yacheiki*) consisted mostly of young scientists. With some exceptions, the prestigious research institutes had directors whose prominence in science had already been established during the pre-Revolutionary period. Until the beginning of 1929, the Academy of Sciences of the USSR did not have a single member who was also a member of the Communist Party. Although membership in the Party was not obligatory for scientific promotion, the role and influence of the Party structure *within* scientific units were permanently growing, and the proportion of active Party members among research and administrative personnel was increasing substantially. Today probably more than a half of all full academicians and corresponding members of the Academy of Sciences of the USSR and other academies are members of the Communist Party as well. The same is true of the Party/non-Party ratio among the ranks of professors, heads of laboratories, directors of research institutes, and rectors of higher educational institutes and universities. The role and power of internal Party committees and bureaus are now very real and important factors in the decision-making processes.

In 1971 a special decision of the Central Committee of the Communist Party of the USSR made the committee or bureau Party secretary's position within research institutes and higher schools much more important than it had previously been. (The difference between a Party committee and a bureau is both quantitative and qualitative. A bureau is elected when the institution is comparatively small—about 100–300 staff members. The existence of a committee means that the organization is large and multistructural, with several bureaus in each structure. Committees are normal for universities, academies, and large research institutes with more than 500 staff members. The chairman of the committee could be a professional salaried Party organizer, usually with some scientific background.) The influence of the Party committee or bureau chairman was rather substantial even before 1970–1971, but it was largely a question of *influence,* rather than of real power. In principle, the director and academic council were responsible for appointments and elections, and this authority remained unchallenged when it involved, for example, the question of appointment at the level of senior research scientist or professor. The positions of heads of laboratories and chairmen of departments in universities depended on a secret ballot of the academic council (usually consisting of other professors and heads of laboratories and departments), but the final appointment for the administrative job had to be made by the decision of the presidiums of the academy or of the ministers of education, or health, etc., depending upon the structure of subordination. The Party committee chairmen could exercise their influence only through scientific channels—selection committees or academic councils in the presidium's ministry collegiums.

Since 1971 procedures have been changed, and all prominent research positions, especially in provincial institutes and universities, have to be confirmed additionally by the decisions of local (district, regional, or city) Party committees. At this level the opinions of the Party chairman and the Party bureau are more decisive than those of the director or academic council. The administration and academic councils still largely control junior research positions, but all senior grades have become much more dependent on the political system. This requirement of double confirmation for all senior appointments also makes it easier for the KGB to interfere in the process of scientific promotion. It is therefore not difficult to understand why the situation is now really very different from that of the pre-1971 period.

Neither directors of research institutes nor deans or rectors of universities are permitted to communicate directly with the KGB's central or local system in order to obtain access to an individual's file. These positions are administrative, not political, and in many cases they are occupied by prominent scientific figures who are not always members of the Party. Nor is the chairman of the university Party committee able to ask for somebody's secret KGB file. However, he (or she), as well as the director, of course, could, if they liked, ask the KGB's advice through the "special

department" in their own administrative system. The same restrictions apply even for academy presidiums. They are composed of "elected" scientists, and they are not entitled to investigate the real power structure of the single-party state, or,
indeed, secret files on their colleagues. However, at the level of
the regional and local Party chairman, the situation is already different. Persons in these positions can be briefed by KGB officials
about the more hidden aspects of a person's life, and may read
the files when they become available in the local KGB office (or
receive them through the KGB from outside sources in the case
of persons who still live elsewhere). The heads of the local KGB
offices are usually members of the regional Party committee, and
(since Khrushchev's time) they are under a system of double
control, supervised by both central KGB headquarters and the
regional Party chairman. (During Stalin's time the security system was independent of the local Party system. Khrushchev gave
the Party apparatus more effective priorities in the power structure.)

The new situation made it practically impossible for a number
of dissident scientists, forced to resign from their own institutes
or universities, to find equal or higher positions in a different administrative hierarchy (e.g., from the Academy of Sciences of the
USSR to the Academy of Medical Sciences or to a research institute of some ministry or other).

Through the new arrangements for senior appointments, all
previously unrelated research networks differentiated according
to professional specialization became connected because of their
general links with the universal Party system. The resulting increase in the possibility of restrictive measures can be illustrated
by several examples.

In 1962, I decided to resign my senior research position at the
Moscow Agricultural Academy (a kind of agricultural university)
because of political pressure over my *samizdat* work about Lysenko. However, I quickly found a new, more senior post as the
head of the Laboratory of Molecular Radiobiology in the Re-

search Institute of Medical Radiology, which was within the Academy of Medical Sciences of the USSR and situated outside of Moscow (in the Kaluga region). Nobody there was aware of my Moscow story. When I was dismissed in 1969 after more serious dissident political activity (the new *samizdat* book on international cooperation in science as well as unofficial publication abroad of a book called *The Rise and Fall of T. D. Lysenko*), I was unable to find any professional job for almost two years. The reason for my leaving Moscow was indicated in my "employment book" (a document obligatory for all persons stating the sequence of employment) as "voluntary resignation." The new entry was "dismissal for political reasons," and any new employer would be well aware of what this meant. Eventually I was able to obtain a position as senior research scientist in biochemistry at an obscure research institute in the small provincial town of Borovsk. At this time, dissident scientists with a more or less good research reputation were able to find work in their profession but at a lower status.

My case was not exceptional. I could describe many identical ones—dissident scientists who were not well known because their political activity had not been related to publication abroad. There were, for example, members of the human rights movement whose signatures were among hundreds of others protesting against various political trials. The action itself (signing a protest, etc.) might have taken place in 1968 or 1969, but demotion often occurred later, particularly if the individual involved refused to "repent" or criticize past "mistakes." Here are just a few examples to illustrate the new system in action:

N., a rather prominent and internationally well-known biologist, was a professor at Moscow University and head of the laboratory in a research institute within the Academy of Sciences of the USSR. However, he signed various political protests in 1968–1969. Two years later he lost his professorship at the university. After two more years, he was not confirmed as the head of his own laboratory (these positions are subject to "re-election" every five years), but he was left as a senior research worker at the same institute. This demotion was in no way related to N.'s academic performance. In fact, his research ability was at its prime, as was quite obvious from the recognition his publications received outside the USSR.

T., a prominent applied physicist, wrote a *samizdat* essay on social questions which had a restricted circulation but was never published either in the USSR or abroad. However, the research institute where he was a senior research worker, and where he enjoyed great respect, was engaged in space research and other classified work. Thus T.'s resignation, after several reprimands, was inevitable. However, as a doctor of science, he easily found another higher and better-paid position at a research institute within the industrial scientific network, independent of the academy structure, where doctors of science were comparatively rare among staff members. T. engaged in successful work as the acting head of a large department, and his new popular scientific book was accepted for publication. At the same time, he continued his dissident activities, but outside of his research institute, and the administration of his institute knew nothing about it.

T.'s appointment as acting head of the department had to be reconfirmed through "election," and at this stage his dissidence somehow became known to the administration. T. was not re-elected but was offered the possibility of remaining at the senior research level—without administrative duties. (It is impossible to offer a junior research position to a doctor of science; thus degrees and titles have some value even for dissidents.) T., however, decided to resign and openly protest against this injustice. But his disobedience had just the reverse effect. The publisher did not publish his book, or return the manuscript, but just postponed publication indefinitely. (Publication would have meant publicity. From 1970 censors had begun to study not only manuscripts as such, but also the files of authors.) T. chose open resistance and more overt political activity and now was unable to find professional work even at a level which would have been available to him a short time before. He has been unemployed for more than three years, in spite of the high demand everywhere for experts with his qualifications (in computers). He probably would be able to find a senior research position in an obscure provincial research institute, but he has not yet reached the stage where the basic necessities of life force him to leave his privileged Moscow residence.*

The third example comes from the social sciences, where political and ideological control are much more rigid and where scholars are considered to be less relevant to the "productive forces" and are therefore more easily made "redundant." In this case, I can provide the name of the scientist concerned, because he himself made an appeal for support in the form of "An Open Letter to Colleagues," published recently in the United States. I quote from the letter of Dr. I. A. Mel'cuk, described by the American colleague who submitted the "Open Letter" for publication as "a very distinguished linguist, certainly one of the foremost scholars in this field."

From "An Open Letter to Colleagues"
by I. A. Mel'cuk

On March 15, 1976, at a session of the Faculty Board (*Ucheny Sovet*) of the Institute of Linguistics, Soviet Academy of Sciences, Moscow, I was not re-elected senior research fellow, the position I held up to that time. According to the regulations senior research fellows must be re-elected by their Faculty Board every five years. It is usually a routine procedure which, as a rule, the person concerned should not even attend. If he is not re-elected, the researcher must be fired by the administration no more than a year after the decision by the Faculty Board took place.

I have been with the Institute of Linguistics since 1956,

---

* In 1977 T. emigrated from the USSR. He now lives and works in the United States.

and have written and published more than 150 linguistic works, including several books; many of my papers are translated and published in the United States, France, Spain, West Germany, Poland, Hungary, and East Germany. I have often been invited to take part in international linguistic conventions, to give lectures, to serve on editorial boards of Western linguistic periodicals, etc.

The only reason explicitly stated for dismissing me was my letter published in the *New York Times* on January 26, 1976. This expressed my disagreement with the campaign waged against Andrei D. Sakharov by the Soviet press, as well as my protest against the arrest and the trial of an eminent Soviet scientist, the biophysicist Sergei Kovalev, who has been sentenced (on purely political charges) to seven years in prison and three years more in exile. . . .

But those speaking at the Faculty Board meeting said, for example, that my letter to the *Times* "besmirches our country and covers with shame any research worker, . . . such an action is inadmissible not only in the scientific community of the Institute, but in the community of all Soviet people as well" (V. Yartseva, the director).

According to another speaker, Mr. Guzman, by "slandering our country I. A. Mel'cuk does serious harm not only to it but also to all progressive mankind . . . so that his hostile action makes impossible the further presence of I. Mel'cuk in the Institute of Linguistics . . ."

I was and am forbidden to teach, to take part in many scientific conventions, to go abroad to meet Western colleagues. Immediately after the appearance of my letter in the *New York Times,* Soviet linguistic periodicals and publishing houses began suppressing references to my works, as well as acknowledgments by other authors mentioning my name; one publishing house (Progress) even suppressed my name as an editor or translator.

Under such conditions, I can no longer carry out regular linguistic research. Which amounts to presenting me with a tragic choice: either a meaningless existence in my country . . . or emigration—a lifelong separation (such is Soviet law) from my native land as well as from my relatives, friends, and colleagues [*New York Review of Books*, Oct. 14, 1976].*

It would take too long to discuss the functioning of the new system of political control in all areas of scientific life, but certain main points should be noted. For the most prestigious corresponding members of the academies and full academicians, final decision is still by secret ballot of a general meeting of the academy. However, potential candidates go through a preliminary screening process not only in the Party committees of the academies, but also in the Science-Education Department of the Party Central Committee (or republic committees, if it is a question of the republic academies—Ukrainian, Armenian, etc.). Academicians who are Party members are informed of the Party recommendations *before* the election takes place, and they must vote in accordance with Party discipline. There have been occasional instances of disobedience, but most of them occurred five or six years ago. Now, with nearly half of the academicians within the Party ranks, it would be rather surprising to have a secret ballot with unexpected results.

Not only is the awarding of the Lenin Prize, or other special academic prizes, subject to preliminary consideration at the different Party levels, but the nomination of candidates for such prizes (the names of candidates are published in the press, and this is already a mark of distinction) must be accompanied by a special character reference and discussed at the Party bureau and signed by the Party chairman *before* it is signed by the director of the research establishment. While the positions of senior research worker or professor are subject to confirmation by the regional or city Party committees, the positions of directors of research institutes, rectors of universities and higher schools, and heads of special research projects are subject to confirmation and approval by the Party Central Committee secretariat as well

*In 1977 I. A. Mel'cuk emigrated. He is now a professor at the University of Montreal.

(or, where relevant, by the secretariats of the republic Communist parties, i.e., when the institute comes under the jurisdiction of the academy or some other system at the union-republican level—Georgian, Ukrainian, Tadzhik, etc.). The presidents of the academies, who theoretically, in accordance with the bylaws, must be and are in fact "elected" by a general meeting of their academy every five years, are recommended for such election after discussion at the Politburo level. At the last such event at the Academy of Sciences of the USSR, the recommendation to elect the prominent physicist A. P. Alexandrov was made personally by Politburo member N. A. Suslov at the Academy general meeting, although Suslov has no direct administrative connection with the Academy system as such.

The Party not only controls the selection of scientists for exchange programs or for participation in meetings abroad; it also screens Soviet speakers for international congresses held within the USSR. Dissident scientists who are often able to take part in Soviet internal research conferences do not find it easy to attend international meetings taking place in the USSR. Occasionally, when the selection of the main speakers at an international congress is the responsibility of an organizing committee of an international association working abroad (rather than individual national organizing committees), it can happen that prominent but politically undesirable Soviet scientists are chosen. However, in such cases, they either are prevented from accepting the invitation or find it difficult to get clearance for their papers in advance (papers delivered at international congresses held in the USSR must be censored before presentation). Simply arriving at the congress could mean inviting trouble. I have had personal experience of this kind. When I was invited by the International Association of Gerontology to give a lecture at the Ninth Congress on Gerontology in Kiev in 1972 (it is considered to be an honor; also, lectures, as opposed to the presentation of experimental work, can be given without special permission from the research institute where the lecturer works), I was simply arrested in Kiev by police near the Congress Hall, detained for several hours, and then deported to Moscow, with a warning not to return.

This was an absolutely illegal action, but the Kiev KGB officials were probably unable to find a legal means of implementing some "recommendation" from above that measures be taken to reduce my stature in the eyes of foreign gerontologists. At this time I held an official position as senior research scientist, and my research was directly related to problems of aging. I was also a member of the Gerontological Society of the USSR and was eligible to attend the congress in Kiev, and most probably would have been able to do so as an ordinary member of the audience had I not been invited to give a lecture. I exercised this right simply to *attend* the International Congress on Biophysics one month later in Moscow.

Publication facilities are important for recognition and promotion. Dissident scientists working in minor research positions in the natural or technical sciences can get their papers published in academic journals, but their *books* are rejected or suffer endless delays in the process of consideration even when of a very high standard. They are not able to publish more popular scientific articles in magazines with wide circulation. Publication of an experimental paper is considered normal, but the writing of a monograph or popular science book is regarded as a privilege of prominent and distinguished scientists. The book is confirmation of the prominence of the author. In the 1960s, the manuscript of a book would be carefully considered, reported on, censored, and edited if necessary. The author had only to produce the document confirming his employment within the appropriate research system; his personal files would not be required for a decision about the publication. However, since 1971, personal files on the author can also be demanded either by the editor in chief, the director of the publishing house, or the censor attached to the relevant field of publication. Heads of publishing houses as well as senior censors are automatically entitled to have access (through the KGB system) to classified information.

This differentiation between categories strictly academic,

scholarly, and general is achieved through different levels of blacklisting. Censors in charge of examining foreign literature which arrives by post for libraries and individuals are now specialized (as I discovered when investigating the distribution of their personal numbers on books and journals [9]). Censors of domestic literature are, however, highly specialized and have different qualifications and instructions depending upon their professional expertise. Categories of secret information in medicine are not necessarily available for a censor working in a publishing house devoted to technical literature on cars or the railroad industry, although every censor has a general blacklist of names which must not be quoted or mentioned without special permission (e.g., Trotsky, Khrushchev, Bukharin, etc.). However, censors of specialized academic journals do not have additional long lists of politically dubious persons who, nevertheless, are working within the Soviet scientific establishment. For example, there is the case of N. V. Timofeev-Resovsky, who worked for a long time in prewar Germany, and was captured by the Russian army, served about ten years in a prison research institute, and was never officially rehabilitated. However, his professional papers and several books were published during 1957–1971. His name is often mentioned in academic publications dealing with his special field, where he is a real authority, and many references even refer to his early research published in Germany, because it has now become classic. When, however, he wrote an article for a more widely circulated popular science journal, it was turned down. And when his name was quoted in an article by another author, written for the central national press, the name was removed before publication. The editor later reluctantly explained that this was not his own action but that of the "editor of Glavlit" (the official term for the censor). Timofeev-Resovsky's name was not on a blacklist for academic literature, although it was forbidden for more general publication, as the ordinary public can only be allowed to know about the achievements of *good* Soviet scientists.

Provincial censors certainly are not as well informed as centrally located publishing organizations, and probably that is why provincial newspapers are not available to foreign readers in libraries. Similarly, it is not possible for libraries abroad to subscribe to local (regional or city) Soviet newspapers.

The increase in political control over recognition and promotion in science evident during the last years does not mean that there has also been interference with the substance or direction of scientific research (i.e., in nonclassified areas). I can see no indication that the system of planning, funding, or final results has really changed in science. Certainly there has been an improvement in the quality of scientific and technical work, which is in no way the result of political control. It is simply the most obvious result of the end of duplication as an approach. It is also related to better international coordination and cooperation in science, with the effect that new equipment, methods, and facilities have become available. State interference with the substance of research did occur, but it was not in fact political; rather it was a question of preferential support for new, more modern, and original ideas. The projects which resulted from these policies were created by the advanced research communities themselves. The government's recent statement, for example, concerning special organizational and financial support for molecular biology and molecular genetics was the result of many discussions within scientific circles about their needs and expectations. The pseudo-scientific practices which dominated many fields in the Stalin era, and partly survived during Khrushchev's time, did not disappear overnight after Khrushchev's fall. However, by 1971–1972 they virtually ceased to exist in the natural sciences. The new younger generation of research workers had much higher qualifications and better contact with foreign colleagues. I do not believe, however, that special decisions play a major role in the general functioning of science in the USSR. The interests of the state, now that the USSR has become a real superpower, demand excellence in science and effective technology. It was finally acknowledged that this could be achieved through a process of natural integration, by removing the walls isolating Soviet science from the more advanced science of the West. Inevitably, the result has been an increase in the importance, prestige, and influence of science within the country as well as an extension of its value for the state. At this stage, political control is a question of attempting to shape the ideological attitudes of the new more influential scientific and technological elite. The state grants more freedom and provides better facilities for *useful* research, but at the same time it is determined to claim that scientific progress is proof of the advantageous and progressive nature of the socialist system and socialist ideas. This is why research workers are now given much more *scientific* freedom, while political freedom remains extremely limited. Many good scientists are aware of these restrictions, but adapt to the situation for the sake of scientific advance. Some do not accept this compromise, and pay a price for their dissent. However, there are also a large number of scientists who do not think in these terms at all and do not regard it as a compromise. They genuinely believe in Communist ideals, although not all of them associate these ideals with the bureaucratic character of the Soviet state. Some would prefer a more liberal kind of socialist democracy with greater freedom and respect for the rights of the individual (i.e., there are also hard-liners and liberals among the scientific elite). It has been Party and state policy since the Revolution to create a powerful and productive science and technology. In the past unfortunate and misdirected efforts often had a negative effect. The intention was to create a unique *Soviet science,* a new miracle of human activity, giving new answers and new *socialist* solutions to scientific and technological problems. The results of this program were expensive and often unsuccessful—in many cases an international scandal. This painful experience, against a background of general political change, finally taught the ruling group that problems such as protein biosynthesis, inheritance, or nuclear fission could be solved in both the USSR and the United States by the same methods. The myth of the peculiar advantages of a unique *Soviet* science is now almost dead. The Soviet government does not, however, favor complete scientific and technological integration with the West, which is now beginning to take place. Its policy is to maintain the existence of the *Soviet scientific community* as a separate and distinct entity within the world scientific community. Although recent trends toward integration reflect a largely pragmatic approach, essential characteristics of the system inevitably influence the final result.

Until recently technological backwardness and isolation from the rest of the world prevented Soviet scientists from receiving what they required for really successful work. Today the picture is different; the state really does provide greater technical assistance, and has brought about controlled but nevertheless fruitful access to foreign research facilities.

My own views on this question were formed during 1967–1968 (see my book on international cooperation in science published in 1971 [9]), and are rather different from the official approach. I believe that science and scientists must be free from this kind of political orientation and control in the USSR and elsewhere; this will not only encourage scientific cooperation but also extend the political influence of increasingly significant universal scientific communities, which will in turn reduce the level of antagonism between different political systems and make the world a less dangerous place to live in.

The development of technology, the advance of science, the quest for knowledge—these are universal and noble pursuits; but political abuse can easily channel them into competitive, dangerous, and destructive directions.

TREATMENT OF DISSIDENT JEWISH SCIENTISTS: A SPECIAL CASE

The various forms of political control on Soviet science* that have already been described scarcely apply to the Jewish scien-

---

* I use the word "science" in the Western sense of the word, for experimental work in physics, chemistry, and biology.

tists and technologists who show their disagreement with the Soviet system by applying for permission to emigrate to Israel—the tolerance is far less. Such would-be emigrants may be demoted from higher and more responsible positions, but dismissal from any kind of research post is common, although it is less frequent for those working in medicine, psychology, and the humanities. No comprehensive figures are available; my impressions are based on cases I have personally known of or read about in literature dealing with the special plight of Jews in Russia.

Several chemists and physicists whom I knew were dismissed when they applied for emigration, and those who had worked in classified areas of research (and almost all areas of chemistry and physics are classified) had to spend several years of "cooling" without professional employment. Two senior historians and two well-known psychologists, on the other hand, were allowed to continue working at their institutes until just before the arrival of the emigration papers, and the same procedure was followed in the case of several medical practitioners.

I do not know of any official ruling behind these differences, but my guess is that the decisions are made on pragmatic and economic grounds. The value of a doctor's work is, for example, the same after he has applied for emigration as it was before. Historians and psychologists do not need much state funding, but chemists and physicists do, and the authorities are therefore unwilling to finance projects in which potential emigrants are concerned. In addition, experimental scientists may lose their high qualifications while they are unemployed more quickly than do the historians, social scientists, schoolteachers, and others. Scholars in the humanities at least have time after dismissal to prepare themselves for adapting to a new life and to study a new language, especially if they are helped to a certain extent by Jewish organizations abroad.

I may be crediting the official line with more thought than it has actually taken, but my impression is that as soon as worldwide pressure and internal discontent made emigration inevitable, instructions were designed to prevent turning Russia's brain drain into a blood transfusion for Israel or elsewhere. "If we cannot prevent emigration"—I imagine someone briefed the emigration officials—"let us shape the trend to create as many problems, embarrassments, and disappointments as possible for the countries receiving the emigrants."

The whole arrangement is a strictly one-way road of *deportation*, along which the emigrants are stripped of their original citizenship and are prevented from returning to the country of their birth. As in all deportations, the process is designed to get rid of those who are unwelcome, troublesome, or irritating, and to create more problems for the deportees and their host country than for their homeland, which dismisses them and deprives them of the normal and legal rights of return.

Hypertrophy: The Latest Development in Soviet Science

The high rate of development of Soviet science and technology (from the quantitative point of view) during 1918–1971 had many facets. One of the most important of these, as I have already indicated in the previous chapters, was the principle of self-sufficiency, the notion of the independence of *Soviet* (or socialist) science and technology, which must be competitive with *all* important projects and problems under development in foreign countries and able to produce some unique "socialist" solutions. During Khrushchev's period and for a few years afterwards, this impossible task was simplified by giving permission to duplicate foreign achievements and models when the superiority of certain branches of knowledge abroad was evident. However, either with permission to duplicate or with the requirement of more independent solutions, *all* branches of knowledge had to be represented within the USSR. This was considered to be a strategic necessity.

Such a policy certainly led to the creation of an enormously increased scientific network. The USSR became the country with the largest maximal number of research workers per million inhabitants. There is no country in the most advanced parts of the world which could afford such a high proportion of professionally qualified research workers and research assistants and so many different research establishments.

Potential redundancy was also stimulated by the policy of national equality. All fifteen union national republics are formally equal, and when the biggest of them (Ukraine and Belorussia) established their own national academies of sciences, it was inevitable for others to follow the trend and to establish their own academies as well. All these academies organized their structure according to certain standards, with presidiums, divisions, and research institutes in all the main fields (physics, biology, chemistry, etc.) and with some specialized local research units. All academies started to publish their *Proceedings* and other journals. The Soviet Union as a state does not need all these provincial academies; there are too many of them, and duplication or low-quality research is inevitable given such structures. Each academy has its own bureaucratic apparatus, with planning and coordinating commissions and much more. In many union republics there were not enough scientists with high qualifications to be elected as academicians, and in many cases provincial "national" academicians are very mediocre research workers, not well known within their professional fields. The establishment of so many academies was a matter of prestige rather than necessity, a symbol of the equality of national minorities. This multi-parallel structure does not really contribute to the best scientific development of Soviet science as a whole. The trend to create the provincial academies, which was finally completed with academies of sciences in all fifteen union republics (the Academy of Sciences of the USSR also represents the Russian federation), big or small, induced a chain reaction in specialized academies as well. At the beginning of the 1950s an attempt was made to start the development in the union republics of a chain of republic academies of agricultural and medical sciences. However, this was too much for the country to afford, in terms of financial and human resources alike. Several such academies (the Ukrainian Academy of Agricultural Sciences, the Uzbek Academy of Agricultural Sciences, etc.) were abolished in 1962. Research workers who had acquired the title of academician and special supplementary salaries simply lost them both.

The hypertrophic growth of Soviet science was not the result of these two factors only. The simple fact that the research profession has become the best-paid and most prestigious occupation had its own impact. Every branch of industry, every professional field, wanted to establish its own research network. The system of grants does not exist in the USSR. Therefore, it is often easier for one or another All-Union or republic ministry to establish a new special research unit than to give a grant to a research group already existing somewhere for the development of a certain project. Temporary establishments are neither popular nor easy to create; good research workers want more permanent positions.

In 1970 the government decided to reduce the number of independent research units (laboratories, departments, etc.) throughout the country and in this way to reduce the budget for science. The salaries of some research administrators are much higher than the salaries of senior or junior research workers who do not hold administrative posts. To attract the more prominent scientists from one institute to another, it was often necessary to offer them a higher position (with a higher income as well). Senior scientists might receive offers to be heads of laboratories in other institutes, even when these laboratories did not really exist, or had only two or three research workers. To prevent this practice of creating "micro-laboratories," the decision stipulated that the name "laboratory" could be given only to a group of at least twelve persons: five to six research workers and five to six technicians, engineers, and other supplementary staff members. Minimal limits were laid down for "departments" (at least three or four laboratories), "divisions," and "research institutes" as well. If an existing laboratory was too small, it had to be reduced to the status of "group," and the post of "head of laboratory" would then disappear, with a substantial saving for the research budget.

However, the government's good intention produced precisely the opposite result. In most cases the small laboratories and departments made all possible efforts to become larger so as to be qualified for the name they bore. The academies and other research institutions turned a blind eye to this misrepresentation of the government's decision.

I shall not try to list the many other factors which stimulated "growth," but it became obvious that the new *détente* policy made many scientific units potentially redundant. Many research institutes, laboratories, groups, and design bureaus which had been created either in the period of "universal self-sufficiency" or in that of "duplication" became useless as soon as acceptance of the integration principle and the purchase of new foreign technology, or licenses for production of this technology, became an irreversible trend in the state planning program. Many internal agreements for research on certain groups of industrial or consumer products, and many government orders which had been placed within the USSR, were simply canceled, because the designs, equipment, chemical products, and some consumer products could be more easily, cheaply, and quickly obtained through trade and cooperation.

Valid statistics are not available, but it is clear from the new designs and inventions registered in the USSR that after 1970 their number started to decline with accelerating speed (58). It can be assumed that the reduction in the number of research establishments and research posts should have the same trend. But the reduction of something already existing is a more difficult problem to a bureaucracy than its increase and expansion. It became much easier, with the excuse of potential redundancy, to dismiss dissenters within scientific establishments or to encourage the emigration of Jewish research workers when they were not considered important.

It was necessary, nevertheless, to find appropriate occupation for latently redundant parts of research and scientific facilities in both academy and industrial research networks. The only reasonable solution has slowly started to take shape in recent times as a logical result of many independent factors. One of them was the permanent appeal and demand from the state to orient scientific research more specifically toward current practical problems. Science is not just a search for an increase of knowledge; it is also a direct productive force. This was the official line for many years, and many research institutes which had disregarded it earlier could no longer ignore it in the new situation.

Another factor was directly related to the usually very good workshops and technical facilities attached to many research institutes to enable them to study and duplicate specialized pieces of foreign equipment during the "duplication" period in the development of Soviet science. These often excellent workshops, with good engineers, technological experts, and a highly qualified working force oriented toward high-quality personal craftsmanship, were not fully utilized when the duplication trend came to its conclusion. For example, the excellent and spacious workshop, a kind of small plant projected and built as part of the Research Institute of Medical Radiology in Obninsk (where I worked in 1963–1969), was initially oriented toward evaluation, duplication, and final synthetic design of radiological and X-ray equipment on the basis of German, French, Japanese, and other models. This objective was reduced or canceled after 1970–1971, and only technical testing of foreign equipment was left as the main task. However, to keep craftsmen and technical staff usefully occupied, this workshop (as well as many others in different institutes) started to produce commercially viable pieces of laboratory equipment on the basis of individual orders from laboratories, other institutes, and educational institutions. Some designs became much more popular and of better quality than industrial commercial products, because they had been made with more care and by workers of higher qualification. They also had been made not just for general consumer use, but for a special contractor, who could complain if something were wrong. Many other pieces of equipment were just not available as commercial products or were made according to individual specifications.

At many technical exhibitions the products of small workshops were often evaluated higher than general industrial products of the same kind. Even in such fields as nuclear research equipment the best Soviet-made mica-window and windowless "gas-flowing" Geiger counters were produced in the workshop of the K. A. Timiriazev Agricultural Academy. This small workshop was under strong pressure from many research and industrial units to sell these counters, because they were much better and more reliable than mass industrial products of the same design.

Two talented workers, real craftsmen, were responsible for this success; they discovered a way to make mica windows about five times thinner than general industry found it possible to do. They also tested each counter they made to be sure about its quality. Mass industrial products are tested only selectively. Small workshops started to be competitive with industry and often won, not only in the internal market but also for export to Comecon countries. It became usual for the laboratories which install equipment for radioactive measurements to attach mica-window workshop-made Geiger counters instead of those which were supplied for the other facilities.

Highly centralized state-owned industry can often successfully produce pieces of equipment for heavy plants, or for mass state and private consumers (tractors, cars, machines, refrigerators, drilling equipment, ships, etc.). However, it is not well adapted to the requirements of small-scale and individual consumers and for things which must be made in small numbers. This is true for chemical and biochemical products as well. State-owned industry produces a lot of high-quality sulfuric acid, chloric acid, explosives, fertilizers, and so on, but when an order is received for enzymes, hormones, special proteins, nucleic acids, amino acids, and other biochemicals which are needed in small amounts for a few laboratories, the industrial ministries show little interest. And as the story of the mica-window Geiger counters indicates, industrial products of this kind are so poor in quality that they are simply useless, and the parallel line of "self-made" production within the scientific establishment becomes necessary. It took some time to accept this initially spontaneous process as normal and to include it in the central and local planning arrangements. Research institutes started to receive orders from the planning system and to produce commercial materials and equipment when *small-scale need for highly reliable products* was demonstrated. Electrophoretic apparatus, column chromatography systems, complex laboratory glass designs, many enzymes, many isotope-labeled biochemicals, some nutritional products of bacteriological origin (lysine, tryptophan), and countless other products began to be commercially manufactured by research institutes and laboratories. Science became a productive force not in an intermediate but in a direct sense. Even highly prestigious research institutes of the Academy of Sciences of the USSR started to work in the same direction. The Institutes of Genetics started practical work in plant and animal breeding and selection, the Institute of Biochemistry became the main adviser and developer of fermentation processes in the tea, tobacco, and wine industries, and the Institute of Organic Chemistry, headed by the former president of the Academy, A. N. Nesmeyanov, announced a revolutionary idea—to develop the commercial production of artificial beluga black caviar—the most expensive food product in the world. During the last two decades natural caviar has become a rarity because the only source of beluga fish—the Caspian Sea and the big Caspian basin rivers, the Volga and the Ural—has become so polluted that the fish population has been seriously depleted. According to the research program, an almost identical caviar could be made from egg proteins with special supplements and hydrophysical and membrane technology. In 1973 when I took my leave for a year of research work in London, A. N. Nesmeyanov's institute had started to build a new wing for the production of commercially viable synthetic beluga caviar. My stay in London became much longer than I expected, and I have no current news about this new wing. But I think that it was completed long ago. I have not been able to find the new synthetic caviar in the British food shops (they advertise and sell at a very high price the *genuine* Caspian caviar from the USSR and from Iran). I am told the synthetic caviar is not yet available for Soviet consumers, either. I do not know whether the synthetic caviar was a failure or whether it was such a great success that it is really impossible now to distinguish it from the traditional natural product.

Chapter 9

Some Forecasts

The picture of the development of Soviet science and technology during different periods of history of the USSR shows clearly that it was rather uneven, contradictory, in many cases misdirected, and beset with pseudo-scientific phenomena. These aberrations resulted from political experimentation, dictatorship, incompetent leaders, ideological interference, cold war, and many other factors mainly from outside the scientific community. Despite human loss, isolation, and pressure of every kind, at the heart of Soviet science there were always a number of brilliant and prominent figures of high principles and integrity, who were able to carry on in the best tradition of Russian and world science.

During sixty years of a new social system science and scientists have provided the main links between Soviet society and the outside world because only for science were these links absolutely essential for survival. It is quite clear that these links will be more important in the future and scientific cooperation will provide the basis for cooperation at all other levels. Science, besides the part it played in the internal development of a very large, mostly self-sufficient, and very suspicious country, represented its eyes, ears, and all other senses in its interaction with the outside world, and explained it to other population groups, including the ruling political elite. This social function of science within a semi-closed society is extremely important, and it will become more so when there is better international cooperation.

The general forecast which I can provide for the reader is optimistic, if considered from the point of view of those who would like to see the USSR as a peaceful member of the international community. It is not, however, optimistic for those who expect that the USSR will finally collapse or disintegrate, become bankrupt, or somehow disappear as a world power. These expectations, which once made the gloomy predictions in A. Amalrik's book *Will the Soviet Union Survive until 1984?* so popular, certainly never proved to be true. Therefore it is better to face reality.

Considering the possible ways in which Soviet science and technology will develop in the future, it is convenient to isolate different problems and examine them separately.

Scientific Freedom and International Cooperation

The whole problem of scientific freedom is international and Soviet science cannot be isolated from worldwide discussion on scientific freedom and responsibility. Science and technology have become so powerful and sophisticated that they interfere with human life in all developed countries and this interference very often induces more concern than appreciation. The ecological movements (Friends of the Earth, and others), the resistance to the spread of nuclear power stations, the opposition to the use of animals for experimental research, the concern about the possible consequences of genetic engineering—public reaction in these and other areas has become a permanent factor in Western scientific life, and will certainly be increasingly influential in the future. All the problems which face Western research are well reflected in a recent report of the Committee on Scientific Freedom and Responsibility of the American Association for the Advancement of Science (67) and in many other documents. They clearly reflect the general feeling that now that science as a whole has the potential to control almost all aspects of human life, more rigid rules and restrictions must be developed against unlimited "scientific freedoms." These pressures would probably have some effect in influencing Western science, where a major part of the research budget depends on grants, private contributions, and general public attitudes. Soviet science, however, would be relatively unaffected by such pressures, since not only is it financed exclusively through the state budget and state industrial systems, but also the lack of freedom of the press and association prevents the general public from knowing the real situation within scientific establishments and hence from organizing movements against some negative features of scientific research (like secrecy, environmental hazards, etc.).

From this point of view, scientists in the USSR are less free to ignore government attitudes but more independent of public opinion. The consequences of this in the future are rather clear—research in the USSR, although a matter of public concern, has more of a chance of proliferating and succeeding than in the democratic countries because it is supported by the government.

International cooperation in Soviet science is growing slowly along with improvements in the quality and growth of general economic standards. The restrictions on academic exchanges are partly related to bureaucratic and ideological factors and partly to economic and qualitative factors. Political considerations in the selection process for academic exchanges do not overshadow other considerations. For example, a good knowledge of a foreign language and good standing within the scientific field are also important. The time when senior research workers could go abroad with personal interpreters has passed. Everybody who now makes an application for a foreign trip for research purposes must pass a foreign language examination. In this respect, academic exchange has improved and the narrowing of the quality gap will certainly improve international cooperation in the near future.

Soviet insistence on mutual academic exchange agreements is partly related to the shortage of foreign currency funds available for subsidizing the unilateral travel of Soviet scientists and scholars. This factor is temporary, and probably in the future unprogrammed trips from both sides will become more common. Continuing high inflation of almost all "hard" currencies, the sharp decline of all indexes of industrial development, and growing unemployment (including that of intellectual groups) in almost all "rich" countries of the West narrow the gap in living standards in the East and the West. With approaching equality, the economic limitations of exchange will be reduced and the general trend among foreign scientists to spend some of their time (like sabbatical periods) in the USSR will become more common. We can now observe the significant increase of general tourism from the West to the USSR, Bulgaria, Hungary, and other Eastern European countries. This trend is likely to continue. Not only will it improve the exchange of general and scientific information; it will also create new demands and expectations within the USSR, and we will certainly find in the future the expression of dissatisfaction as well. The dissidence in science will remain and probably increase, but the reasons for dissent will be different.

The issues creating dissidence in Soviet science have changed over the course of history in the Soviet Union. The main conflicts before 1964 were related to the struggle between science and pseudo-science and ideological interference in the substance of scientific research. Khrushchev's attempts to solve many serious agricultural and industrial problems without any preliminary studies and consultations also led to conflicts between the academic community and the Party leadership. In 1965–1971 the main issues for dissent were political—the struggle against the rehabilitation of Stalin, protests against political trials, censorship, and so on. During the last few years the right of emigration and foreign travel emerged as the main problem. In the future the issues will also be different, and most probably there will be a strong demand for more general human rights, not by individuals (as at present) but by more influential groups of intellectuals. This trend, even if it does not receive much publicity in the Western press (publicity is always individually oriented), will further the chances of changing the internal situation within the USSR.

The difficulties and at the same time the importance of East-West cooperation in science are well reflected in the recent book by Robert F. Byrnes, *Soviet-American Academic Exchanges, 1958–1975* (68), and I agree with some of his conclusions quoted below:

The academic exchange program has been a relatively constant and stable element in Soviet-American relations since 1958. It began and sustained itself through a time of immense ferment and change within American higher education. It has been important as a symbol. It has served as a link between two contentious states at delicate times, when failure, cancellation, or a breach of one kind or other might have intensified a crisis. It has helped bring about other changes, and it has made prospects for continued peace a little brighter. But the progress has been slight and painful, when one considers the imagination, energy, and resolution expended.

Academic exchanges, and other exchanges as well, increase and improve the Soviet elite's knowledge of the American people, thus beginning to reduce some of the misapprehensions caused in part by Soviet philosophy, propaganda, and way of looking at the world and in part by simple lack of information.

The very existence of the exchange programs and of the more relaxed Soviet-American relationship that they reflect, the increased trickle of information the Soviet government tolerates, and the subtle Western pressure to relax Soviet controls encourage dissidence and dissent among intellectuals, ethnic and religious minorities, and all those who simply want more freedoms. Even the great majority of utterly loyal Soviet citizens have been influenced by the observations and attitudes of their colleagues who have traveled abroad or worked closely with foreign scholars. Thus, every Soviet scholar abroad and every American and other foreign scholar in the Soviet Union helps break down the wall which has isolated that country from the rest of the world [pp. 234–238].

The same conclusions are relevant for Soviet academic exchanges with many democratic countries—Britain, France, Germany, Japan, and others. This trend will certainly continue and its influence will grow. This is my sincere conviction, and this is why I have always argued that the tactic of boycott, which has many advocates in the West, could be absolutely counterproductive. As soon as the connections and the cooperation increase, the mutual interdependence will increase as well, and the possibility of influencing the shape of Soviet internal policy will be more realistic. This conclusion is shared by many Western experts responsible for academic exchange with the USSR. The validity of this attitude is clearly expressed by Byrnes, whose views I have quoted above. Byrnes did not study this problem as a scholar; he was officially responsible for the program of academic exchange with the USSR. Therefore the principle which he expressed is the main principle of the whole idea of scientific cooperation between East and West.

Western governments and peoples have supported academic exchanges, not only because they increase knowledge and understanding, but also because they enlarge Soviet participation in world affairs and open chinks in the wall around the Soviet Union. They see the increasing flow of people and information as a liberating force that may lead to a mellowing and moderating of Soviet policy. In addition, these programs, and their survival already for almost two decades, give us more leverage in negotiating with the Soviet Union than we had, for example, in 1955 [p. 238].

The Quality and Productivity of Research

As soon as there is an optimistic prognosis for the development of international cooperation in Soviet science, it is certain that the quality of research in the USSR will also improve and that it will approach international standards. Even now there are quite a few branches of science where Western scientists working in the USSR on an exchange basis could benefit from Soviet experience. Robert Byrnes acknowledged in his book that American participants in the exchange program benefited significantly in mathematics, basic physics, some branches of medicine (hypertension, immunology, blood diseases), and the Soviet system of medical education and public health service. The advances of Soviet space research made possible joint experiments in space, and while the Americans were more successful in getting information about Mars, Russian space technology and science had the advantage in research on Venus—two soft-landing spacecraft, *Venera* 9 and *10,* provided information which considerably changed many previous ideas about that planet.

With better-organized international exchange and cooperation in research, it is now rather common to see in Western literature articles about Soviet research projects which inspire not only appreciation but sometimes a certain admiration (69, 70). These projects are mostly large-scale installations, like the giant radio telescope in the North Caucasus—the biggest optical telescope in the world, with a neutrino detector, and activated by thermonuclear microexplosion by means of a powerful electron beam—for research in new sources of energy as well as in some other fields.

With the end of the "duplication" psychology and with a completely different system of planning and organizing the scientific program, the clear differentiation of research priorities soon became obvious. In such circumstances cooperation will in the future be as advantageous for Western science as it is advantageous for Russian science now. Many critics of the Soviet system argue that there is absolutely nothing that Americans, for example, can receive from scientific and technological exchange with backward Soviet science. This is not true, because the "backwardness" is not general and the very difference of priorities creates a background for mutual benefit. A recent editorial in *Science* on U.S.–Soviet cooperation (71) also emphasized this point:

> While the Soviet practice is to build scientific research and development into their macroeconomic plans for periods of as much as 15 years, keying it to upwards of 200 priority problems in each 5-year planning segment, their R & D is not locked in so firmly that goals and strategies cannot be changed on short notice. The Soviet passion for planning provides more continuity and stability for science than we do, and their policy recognizes the investment nature of R & D in a way that ours never has. Where we seem to do better is in applying research results under conditions of market choice and risk, even without the help of an explicit emphasis on large doses of R & D in macroeconomic policy. With all the basic differences, both systems seem to produce very good research and innovation.

Scale of Research Projects

During science-propaganda competition Americans proved that they could carry out research programs of larger scale than was possible for Soviet science. The most conspicuous were the Apollo program and several successful expeditions to the moon. However, in general the highly centralized one-party socialist state is better prepared for large-scale research projects than the countries where the governments do not enjoy independence from public opinion, parliaments, press, and different public organizations. Widely publicized large-scale research or technological programs usually evoke strong criticism and public scrutiny and could be an embarrassment for governments. The fate of the Concorde project is a good example of such embarrassment. The supersonic plane is without doubt superb, and it represents a great technological breakthrough; but it does not make a commercial profit and will hardly be commercially viable in the near future. It is enough for the capitalist economies of Britain and France and for their governments to regret having ever undertaken the Concorde design. The same considerations spelt the finish of the British-French undersea channel, on which construction had begun some years before. The Alaska pipeline would probably never have been completed if the oil embargo in 1973 had not changed American attitudes about the environment.

In the Soviet Union large-scale projects (even though commercially unprofitable) are usually exploited to boost government prestige and national prestige as well. The whole economy is centralized, and the general public is not concerned about the profitability of one or another project. When research scientists or industrial workers are sent to the farms to harvest potato fields because of the shortage of manpower in the villages, nobody argues about the price of vegetables. The principle is clear—potatoes (or cabbages, or whatever) are foodstuffs and as such are useful. It cannot be a loss *for the nation*. The same psychology is typical for other projects as well.

The USSR is able to afford large-scale projects not only because it is a socialist centralized state. The country is very big and this determines the large-scale character of many programs. There are plenty of resources available, and large-scale works create many jobs, providing insurance against unemployment. And in each case *the whole nation pays the price*.

Objective experts on Soviet technology acknowledge this potential of the socialist system. Comparing the socialist and capitalist capacities for technological progress, J. Wilczynski in his book *Technology in Comecon* (72), while criticizing the USSR for the current lag in modern technology, organizational defects, and the low quality of consumer production, is, however, optimistic about the future. Asking the question "Which system has a greater capacity for technological progress?" the author gives an answer which I agree with completely, and think it valid not only for industrial development but for scientific development as well:

> There is no simple answer to this question because each system excels in some respects and falls short in others. Socialism has an overall advantage at the macrosocial level, whilst the strength of capitalism lies chiefly at the microeconomic level. Socialism has a great determination and a more effective machinery for extricating an economy from backwardness of stagnation and for shoving it on to a solid road of continuous progress. It can also steer the course of technological development more effectively into the socially most desirable avenues. The economic reforms have decreased or removed several disabilities of the socialist economic system which previously inhibited technological progress of its efficiency.
>
> Capitalism, on the other hand, is more capable of technological improvements at the operational level and along the most efficient lines. It is more flexible and responsive to changing conditions, and modern technology is diffused more widely throughout the economy, so that there are no such glaring differences between different branches as under socialism. Capitalism is in a better position to provide a wider range and a better quality of both goods and services and constantly adapt them to changing needs. Owing to increasing state intervention, some abuses of *laissez-faire* capitalism can be mitigated or prevented [p. 361].

It is easily predictable, for example, that the Soviet space program will keep its large budget increasing for many years to come, while American space research funds are already lessening. However, I do not expect that, in spite of certain improvements in the quality of research, Soviet science will have a lead in such fields as organic chemistry, biochemistry, genetics, electronics, and many others. These fields of knowledge need not only investment, governmental support, and talented scientists, they need absolute freedom of cooperation and exchange of information and the pressure of competiton—conditions which do not exist in the USSR.

Military Technology and Science

It is obvious from the discussion of this aspect of Soviet science and from the above considerations about Soviet capability to carry out large-scale projects efficiently that it will be more and more difficult for the United States and for the West as a whole to maintain superiority if the military race is not restricted by mutually beneficial and reasonable agreements. The socialist state can more easily make military technology a priority area and develop a military-industrial complex of larger scale without any opposition, public protests, interference from press and private companies. Could anybody imagine that the large international airport built near Moscow would be empty and an unoperational "white elephant" for many years simply because some groups of farmers built steel towers near the takeoff and landing runways? This happened in Japan and could probably be expected of Tokyo, but it is unthinkable for the USSR. The United States is vulnerable to the same sort of trouble: the federal government and local authorities have had great difficulty in coming to an agreement about landing rights for the Concorde. The Soviet TU-144 supersonic plane is probably even more noisy, but it now makes flights between Moscow and Alma-Ata. Alma-Ata, as the capital of the Kazakh National Republic, is constitutionally more independent than New York, but would it be reasonable to imagine that the Alma-Ata airport even discussed the question of landing rights? These two examples, while they do not seem scientific, are relevant for military science. It is not pleasant to have a nuclear silo

with an intercontinental missile, or nuclear reactors or plutonium stockpiles, near private residences. This is why Americans must relegate their nuclear and other dangerous materials to unpopulated areas. Russians live near such installations
either without any protest or without any knowledge of them. At
least ten nuclear reactors were operational within the town limits
of Obninsk, where I lived for eleven years, or were situated two to
three kilometers outside. While collecting mushrooms in the
forest, I could often see fenced concrete stores which most probably contained nuclear waste. However, during all my eleven
years there I never heard a single complaint about these "environmental" problems. The logic was simple—if you come to live
and work here, don't worry.

J. Wilczynski (72), in the book from which I earlier quoted,
writes that "socialism has also developed a more effective network and more subtle methods for cultivating and improving the
fighting morale of the armed forces, which represent a definite
advantage in these days of world-wide decline in traditional discipline" (p. 360). This is relevant for science and technology. It is
much easier to recruit bright scientists to work in military projects in the USSR than in most other countries. Up to now the
USSR has proved able to imitate the designs and projects first
developed by the Americans (which is not, however, the case in
consumer-oriented fields of research and development).

Soviet military technology and science has been restricted by
other factors—poor road systems, general industrial lag, time
limits for development of very sophisticated equipment. These
factors, however, are temporary and need not prevail. The USSR,
which is in general much richer in resources, is capable of making unexpected technological advances to become superior in
some fields. It is only a matter of time. The only escape from the
disastrous possibilities of military competition is through reasonable negotiations and the acceleration of world scientific integration which creates interdependence, mutual benefits, and mutual vulnerability.

Theoretical Research

Russian scientists like theories. Theoretical works, however, have also become expensive now. The situation in the current world is such that only big and rich countries can afford theoretical research projects. Applied sciences have become dominant. I can hardly imagine that if somebody in Britain now applied for a million-pound grant and stated that "the work will have purely theoretical importance" the applicant would receive the grant. In the USSR the formula "theoretical importance" is almost magical. It must be included in the proposals for many research programs which must be confirmed by higher officials. Nobody expects that important practical results can be reached without new theories.

Political Control in Science

There is a limit to the demands for conformism and obedience which the Soviet intellectual community will accept, and this has probably already been reached. More pressure and more interference can backfire and induce subtle resistance from a wider group of intellectuals. At the current level of political and ideological pressure, the intellectually bright and thoughtful research workers already have a kind of double life, one real and the other official. A harder line on the human rights issues and on general freedoms will draw more intellectuals into this kind of double life, and probably part of the general population as well. The time of total ideological mass psychosis, typical of Stalin's time in the USSR or of Hitler's dictatorship in Germany, reached the point of no return. The improvement in the material standard of living, when people no longer worry about merely being hungry, could result in the double-life system's creating a situation fruitful for different kinds of underground activities. The strongest political pressure and the most serious repressions against the "democratic" movement and dissidents reached its peak in 1972–1973.

At certain times during these years it would have been possible to create more real and serious opposition if the dissidents themselves had had at least a slight desire for uniting for common aims. Unfortunately, the greatest and most famous figures among random and disorganized dissident groups who could stimulate a serious following—A. Solzhenitsyn, among writers and other humanists, and A. Sakharov, among scientists—suddenly opted for individualistic methods and publicity and proposed a complex of ideas which could not receive serious support from within. Solzhenitsyn appealed to old religious values. Sakharov expressed views about the absolute lack of perspective of socialism as such and about the indisputable advantages of free-enterprise societies. Both these views could only receive publicity and support from outside and, together with other factors, these events finally reduced the dissident activity. However, new waves of dissent can always appear again, and repressive and suppressive measures could stimulate rather than eliminate them.

Science as a Political Factor for the Future

Science, by its nature, represents the most rational force in today's world. It is not yet strong enough to be independent, but this is practically the only worldwide network of organizations and personalities which coexists peacefully and beneficially and freely exchanges available information. Attempts to create global unions of states and governments, workers, writers, and religious or even artistic values, are certain to fail. Selfishness, egocentricity, nationalism, political dogmatism, racialism, and many other dividing factors affect almost all forms of human activity except genuine science. The world scientific community has survived and developed over centuries. Soviet science is not only the instrument of development of Soviet economic and military power, it is also the main channel of knowledge about the outside world and cooperation with the outside world. At this point science and politics interact very closely, and science gains more and more influence. I hope it will be rational, but it must be rational everywhere. For people living in Western Europe this rational approach has begun at a rather strange point. American science and technology promise new neutron bombs which can kill without much destruction of property. Soviet science and technology promise powerful death rays and satellite killers. These two new killers are probably "safer" than old-fashioned thermonuclear bombs with hundreds of megatons of explosive power—they could not only kill people, but destroy everything and contaminate whole countries with deadly radioactive isotopes for hundreds of years. In due course, science will probably find something even more safe for mankind than death rays and neutron bombs. It is possible that in the future, wars will be like games played between sophisticated computers in outer space. But it would be better if computers were used to prevent any form of military conflict.

APPENDICES,
BIBLIOGRAPHY,
AND NOTES

Appendix I

A New Controversy in Soviet Genetics

T. D. Lysenko dominated Soviet genetics and biology for more than twenty-five years. In 1965, when his domination was over, genuine science was quickly restored to research and education. It was generally assumed that his political and ideological interference in the nature of genetic problems had been completely wrong and harmful. The notions of a "Soviet" and a "bourgeois" natural science had been almost forgotten. In spite of Lysenko's long "dictatorship," there were still quite a few prominent, enthusiastic geneticists of the older generation who could build up research again, write new textbooks for schools and universities, and support and teach the younger generation.

For human genetics, however, the position was much more bleak. The subject was declared racialist and was completely destroyed in 1936–1938. Most of its prominent representatives were arrested, and not one of them was found alive when the Gulag camp-prison system was dismantled by Khrushchev in

1965. Details of the tragic fate of Soviet human and medical genetics have been described elsewhere, but the complete liquidation of Soviet human geneticists during the prewar period explains why medical and, in particular, human genetics lagged behind in the rapid development of genetics after 1965.

The problems of human genetics were discussed very briefly, with attention focusing on the pattern of human chromosomes and immunogenetics, in the first Soviet textbook on general genetic written by Professor M. E. Lobashov and published in 1967.[1] In only one chapter did Lobashov really repeat the old clichés on the misuse of human genetics by the bourgeoisie, which through human genetics, he claimed, tried to "justify the exploitation of one class by another, antagonism between different nations and creation of the racialist theories" (p. 679). Lobashov also made some very positive statements about eugenics and considered that previous errors in this branch of human genetics did not necessarily indicate that it could not be developed into a useful field of research. Lobashov's view was that individuals in a human population, although equal in the social sense, have different genotypes and far from identical intellectual possibilities. These inherited differences had to be taken into consideration by educational training programs so that the individual's potential could be released and developed.

At the same time, I. T. Frolov, philosopher and senior official of the Department of Science of the Central Committee of the Communist Party of the USSR, published a small booklet, *Methodological Problems of Genetics*.[2] His main points were designed to prove that the collapse of Lysenko's pseudo-science would not in any way diminish the importance and validity of Marxist dialectic-materialist philosophy for biology and natural sciences in general. Lysenko simply did not use this philosophy correctly; he falsified and distorted both biology and dialectical materialism.

1. M. E. Lobashov, *Genetika* (Leningrad University, 1967).
2. I. T. Frolov, *Metodologicheskie Problemy Genetiki* (Methodological Problems of Genetics) (Moscow: Znanie, 1967).

His theories were "vulgarizations" of the latter; they were "pseudo-dialectic," with many errors of the positivist type.

After his philosophical criticism of Lysenko's ideas, Frolov attempted to present the Marxist background to modern classical genetics, biochemical genetics, and evolution; human genetics was not mentioned in his essay at all, as if it was not part of the problem; the rehabilitation of genetics stopped short, neither ideologists nor scientists being able to analyze human genetics. Nobody wanted merely to acquire knowledge from abroad, but all the roots of Soviet human genetics had been completely destroyed and as yet nothing could appear from the rubble.

The logic of science and the practical needs of medicine, however, made the development of human and medical genetics in the USSR inevitable. The Academy of Medical Sciences started to consider establishing a research institute in the field, and the leading figures of genetics and biology, none of whom had real practical knowledge of human or medical genetics, started discussing its possible task. Initially, the discussions were very peaceful and pragmatic, nobody, quite understandably, wanting to take the lead. Among those taking part, academician Nikolai Dubinin, director of the newly established Institute of General Genetics, was considered to be the main authority. His expertise lay in the field of Drosophila genetics. The next most prominent (but most popular of the geneticists) was academician Boris Astaurov (elected as first president of the Society of Geneticists of the Soviet Union, established in 1966)—an expert on silkworm genetics. He established the Institute of Developmental Biology, which excluded research on human genetics. Of the others actively involved in the discussions, Professor Dmitry Beliaev, director of the Novosibirsk Institute of Cytology and Genetics, had experimented in the genetics of rabbits and some rodents. But the man who was most excited about human genetics, Dr. Vladimir Effroimson, was a silkworm geneticist whose very wide knowledge of medical genetics (he published the first Soviet book on medical genetics in 1963) was purely theoretical.

The other members of the discussions came from similar backgrounds: Professor S. J. Alikhanian was director of the Institute of Bacterial Genetics and Selection; A. A. Prokofieva-Belgovskaya and N. P. Bochkov were cytologists; N. V. Timofeev-Resovsky and R. L. Berg were drosophilists.

Although the project to establish the research institute on human genetics was generally supported, its organization and priorities were not established in the meetings and discussions sponsored either by the Medical Academy or by the newly created All-Union Society of Geneticists. The peaceful but not yet very productive discussions, however, were suddenly sharpened by Dubinin, who, like Frolov one year earlier, published a small popular book under the almost identical title *Some Methodological Problems of Genetics*.[3] The essay strongly politicized the discussion and opened up a new division among geneticists who only recently had represented a strong and friendly joint front against Lysenko.

Dubinin, introduced by the publisher as "a famous geneticist who had great influence on the development of Soviet and world genetics," started his essay in the usual demagogic style of his former Lysenkoist opponents. The methodological principles of the modern natural sciences, he claimed, had been founded by Lenin in his work *Materialism and Empirio-Criticism* (1909), which has eternal importance for the further development of philosophy and all natural sciences, genetics in particular. After a short description of the history of modern genetics (with more than generous attention to Dubinin's own contributions and frequent quotations from Marx, Engels, and Lenin), Dubinin unexpectedly made a statement on human genetics, differing little from Lysenko's approach to the field.

In research on human genetics the main task is to realize that the human race during its development excluded itself

3. N. P. Dubinin, *Nekotorye Metodologicheskie Problemy Genetiki* (Some Methodological Problems of Genetics) (Moscow: Znanie, 1968).

from the evolution of other animals. . . . Now social factors only—class struggle, industrial, cultural, and scientific progress—determine human evolution. Anthropogenesis is completed and ended by the appearance of consciousness. Now social factors are primary and biological factors secondary [p. 60].

At first, this statement did not look out of place. Nobody considered Dubinin an expert on human genetics, and his colleagues and friends tried to explain to him that genetic laws and pinciples apply *equally* to all animals and that man is no exception. Very quickly, however, it became clear that the statement quoted above was not accidental; it was part of a developed system of views which recognized the human race as a *qualitatively* different, new, and final stage of animal evolution.

Two years later Dubinin published another book for the general reader called *Genetics and the Human Future*.[4] By this time his ideas about the qualitative biological gap between man and other animals had developed. He tried to show that genetic makeup in humans bears no relation to the formation of personality. According to Dubinin's theory, human development from birth is determined by "social inheritance," everybody having potentially equal intellectual capabilities. Social inheritance is determined by the character of society itself, and so far only the Communist societies could create the favorable conditions for the full development of human potential.

This book provoked more open discussion, which both sides tried to keep outside serious academic journals (mostly because the discussion was based on opinions, not on facts or research). But Astaurov and Effroimson wrote critical essays which were published in the popular literary magazine *Novy Mir* (vol. 10 [1971]). Frolov, then editor-in-chief of the journal *Voprosy Filosofii* (Problems of Philosophy), organized a "round-table dis-

4. N. P. Dubinin, *Genetika i Budushchee Chelovechestva* (Genetics and the Human Future) (Moscow: Znanie, 1970).

cussion" on human genetics also sponsored by the Advisory Research Council on the Problems of Philosophy of Nature at the Presidium of the Academy of Sciences of the USSR. It was clear from the discussion, which was published in *Voprosy Filosofii* (vol. 9 [1972]), that Dubinin's view received very little support.

The new Institute of Medical Genetics, which was finally established in 1971 within the Academy of Medical Sciences of the USSR, did not have the organization to make a study of all aspects of human genetics possible. It was given narrow objectives and a comparatively young director, Nikolai Bochkov, a medical cytologist. Bochkov's appointment was supported by Dubinin, who expected him to be an ally in the personal confrontations which now made up the discussion.

In 1973, N. P. Dubinin published *Vechnoye Dvizhenie* (Perpetual Motion),[5] a rather self-promoting autobiography. Reviewed very widely and favorably in the official press, it distorted the factual history of Soviet genetics and gave a very mild version of the Lysenko controversy. In his autobiography, however, Dubinin tried to reply to his critics and to formulate the main principles of his vision of human genetics.

> The attempts to whitewash eugenics and to use for this purpose new genetic discoveries had been made by B. L. Astaurov, A. A. Neifakh, and M. D. Golubovsky. V. P. Effroimson developed wrong views about genetic determination of intellectual and personal characteristics of man. This trend started to be an ideological danger. Again alien ideology tried to poison the clear sources of our science. I tried to make uncompromised denunciation of this ideologically harmful attitude. Alas, I met a small but well united group of opponents: B. L. Astaurov, S. M. Gershenson, S. I. Alikhanian, D. K. Beliaev, who defended their wrong position. This worried me and proved that some old geneticists do not understand Marxist ideas about the human race and the prin-

5. N. P. Dubinin, *Vechnoye Dvizhenie* (Perpetual Motion), 2nd ed. (Moscow: Gospolitizdat, 1975).

ciples of society, which is the only factor determining the intellectual specificity of man [p. 434].

He also tried to develop the distinction between the genetic control of physiological and psychological characteristics of human individuals. Genes, according to his view, determine the program of development, life span, some pathologies, expectancy of heart disease, cancer, resistance to infections, and so on. "All these qualities are programmed in the molecules of DNA" (p. 427). The shape of nose and mouth or height are also inherited traits. However, personality and intellect, he claims, depend not on genes, but only on social environment and self-training; their possibilities are unlimited. As soon as man was able to think consciously, he isolated himself from the laws of biology. All "normal" persons are psychologically and mentally equal at birth, and so there is no reason for taking the racialist approach when talking about the *quality* of human genotypes with regard to intellect.

Let us remember the story of the emigration from Russia of many representatives of the bourgeois artistic, scientific and technological elite after the October Revolution. Some eugenicists considered this exodus as an irretrievable loss of valuable genes. But this was nonsense. New talents from the grass roots through Revolution found their way to lead the country, the Party, to create new science, new art, to create the great Soviet Union! . . . This fact illustrates that the creation of personality does not depend on genes, but is induced by the information of social environment [p. 426].

In 1974 academician Boris Astaurov died from a heart attack— a great loss for Soviet biology. His death was probably hastened by strong political pressures induced by the defection of one of the senior research scientists from his institute. Astaurov, as director, had recommended that this good research worker present a paper at an international conference in Italy, but the man asked for political asylum and did not return to the USSR. In such cases, the official bureaucracy makes directors personally responsible. Special Academy commissions started to "investigate" (with active participation by Dubinin) all aspects of the ideological and moral atmosphere of Astaurov's institute. Some senior members of his staff were demoted or dismissed, and Astaurov's health suffered.

Not long before his death, Astaurov, as a member of the editorial board of a popular scientific journal, *Priroda* (Nature), recommended that it publish a translation of the article "Man as a Biological Species," by Ernst Mayr, a prominent American zoologist and population geneticist. The article did not ignore the social aspects of formation of human personality, but it considered *all* characteristics and qualities of individuals to be a result of the interaction of genotype and environmental and social influences.

Mayr's article provoked interesting discussion in *Priroda:* several anthropologists and philosophers supported his views. Dubinin did not join in the discussion in *Priroda,* but instead attacked Mayr's article in a paper published in a journal on philosophy,[6] where he insisted that all the spiritual aspects of human life ae outside biology or genetics. He labeled Mayr's synthetic social and biological approach as "reactionary" and "racialist."

The average time lapse before publication in Soviet journals, including popular scientific ones, is very long, and consequently a reply to Dubinin's article written by D. Beliaev was not published in *Priroda* until a year later.[7] Beliaev, who in 1976 had already been appointed secretary-general of the 1978 International Genetic Congress in Moscow, was reluctant to acknowledge the

6. N. P. Dubinin, "O Filosofskoi Borbe v Biologii" (About Philosophic Struggle in Biology), *Filosofskie Nauki* (Philosophical Sciences), No. 6 (1975), 12–17.

7. D. K. Beliaev, "Problemy Biologii Cheloveka" (Problems of Human Biology), *Priroda* (Nature), Moscow, No. 6 (1976), 26–30.

seriousness of the controversy among the Soviet geneticists. He clearly explained, however, that he disagreed with Dubinin's simplistic explanations. Behavior, he wrote, as well as many of man's intellectual abilities, certainly depends on social factors, teaching, and training, but these factors act on a highly differentiated population and the main organ of mental activity—the human brain—is not excluded from genetic laws and genetic diversity.

Dubinin published a new detailed article on the subject, which appeared in *Voprosy Filosofii* (1977). It mostly repeated his earlier arguments, though it was more politicized. The ideas about the intellectual (or spiritual) aspects of human life became rather mystical, in spite of repeated quotations from Marx, Engels, and Lenin.

> The social life of man is not related to biology. . . . Social inheritance is the only factor in mental development as soon as individuals start to live within society. . . . The qualitative aspects of consciousness do not depend on biology or genetics, they are determined by the participation of man in the social-historic process [p. 47].

> All animal characteristics of human behavior normal for our ancient ancestors are eliminated because they are contradictory of the social position of man as the highest achievement in the evolution of the species [p. 49].

> As soon as evolution created the modern brain (about 30–40 thousand years ago), the new man, *Homo sapiens sapiens*, was set apart from biological evolution. The internal essence of man changed completely, biological laws lost their power over new creation—only social powers became relevant [p. 51].

> Social factors absolutely eliminate the biological [p. 52].

> Social evolution must be ethical and Communist humanism is the only ethical system which represents the ideal of

human activity. . . . The appearance of Communist ethics and morals creates conditions which transform *Homo sapiens* into *Homo sapiens humanis* [p. 54].

The whole article is written in this style. On the influence of animal instincts on human behavior, Dubinin's view is completely opposed to that of Konrad Lorenz. He insists that all animal instincts disappeared in humans as soon as the human brain developed "the second signal system."

Through this complex of pseudo-scientific argumentation, Dubinin is trying to state that the intellectual and ethical progress of *Homo sapiens humanis* is properly directed in one social system only. In any other social system there is no progressive evolution. Social systems create the real man after birth; the priority of genes before birth is not questioned. Good systems create good persons, bad create bad. From Dubinin's autobiography it is clear that he considers himself a good man whose psychology and intellect have been developed under the influence of the best social system. "I am thankful for my fate . . . I do not want to live another life." He is happy. During his seventy years of life, he never discovered that thirty years of Stalinist terror were not good for the creation of *Homo sapiens humanis*.

There is, however, at least one important factor for securing the happiness of *Homo sapiens logicus*—freedom to do research, whether genetic, social, or political. And for the realization of this condition, I do not see much that is ethical or objective in Dubinin's own behavior. The discussion described here is far from just academic. Using his position as the director of the main Institute of General Genetics, his influence as Consultant on Biology for the Party Central Committee, and his membership in different commissions and groups, Dubinin is working hard to suppress by all possible means the development of genuine research in the field of human genetics in the USSR. Even now the Soviet Union is not involved in many important fields of research in human genetics. As part of medical science, human genetics now has modest conditions for research into several specially selected problems, mainly associated with health programs. But as a part of general biology, evolution research, or population genetics, human genetics does not really exist in the USSR.

Appendix II

The Ural Nuclear Disaster of 1957:
Ecological, Genetic, and Population
Researches in the Areas Contaminated by
Long—Lived Products of Radioactive Waste

My article "Two Decades of Dissidence," in *New Scientist* (Nov. 4, 1976), mentioned a nuclear disaster in the South Urals area at the end of 1957 or beginning of 1958. According to my description the disaster resulted from a sudden explosion of the nuclear waste burial ground. Nuclear waste, stored in underground shelters close to the first Soviet military reactors, exploded somehow. Radioactive products mixed together with soil were distributed by strong winds over a large area, probably more than a thousand square miles. Villages and small towns, covered by radioactive dust, were evacuated after some delay. Probably several hundred people died later from radiation sicknes.

I described this tragic accident in an attempt to show that it had made an influential group of Soviet nuclear physicists very sensitive to the biological dangers of radiation. This group became the allies of the suppressed geneticists, who, before this disaster, talked in vain about the mutagenic dangers of radiation, and the genetic consequences of radioactive contamination of the environment for future generations. I was unaware of the fact that this nuclear disaster was absolutely unknown to experts in the West.

My article in *New Scientist* created an unexpected sensation. Reports about this twenty-year-old nuclear disaster were published in almost all the major newspapers (*The Times,* the *Observer,* the *New York Times,* the *Guardian,* and others). At the same time some Western nuclear experts and the chairman of the United Kingdom Atomic Energy Authority, Sir John Hill, tried to dismiss my story as "science fiction," "rubbish," or "a figment of the imagination" (see *The Times,* Nov. 8, 1976).

However, about one month later my story was confirmed by Professor Lev Tumerman, former head of the Biophysics Laboratory in the Institute of Molecular Biology in Moscow, who in 1972 emigrated to Israel. Tumerman visited the area between the two Ural cities Cheliabinsk and Sverdlovsk in 1960. He was able to see that hundreds of square miles of land there had been so heavily contaminated by radioactive wastes that the area was forbidden territory. All the villages and small towns had been destroyed to make the dangerous zone uninhabitable and to prevent the evacuated people from returning. Tumerman's story was published in the *Daily Telegraph* and *The Times* on December 8, 1976, and in many other papers on December 7–9.

Tumerman's eyewitness evidence did not, however, convince some experts, including Sir John Hill, that the accounts of the disaster reported by Tumerman and me were really true. The belief that there was some exaggeration remained, and was expressed in letters by Sir John Hill to *The Times* on December 23, 1976, and February 8, 1977. These doubts made it necessary for me to collect additional valid and easily available information about this nuclear disaster and its real scale.

Different kinds of nuclear accidents release into the environment different kinds of radioactive products. The possible distribution of *reactor nuclear waste* from the waste burial area is rather specific, because the numerous short-lived radioactive isotopes, with very intense gamma and beta radiation, will disappear during the storage period. Only long-lived isotopes, which constitute about 5 to 6 percent of the initial radioactivity, remain dangerous after a two- or three-year period. Chief among these are strontium 90 and cesium 137. Both isotopes have half-lives of about thirty years. Cs^{137} as an isotope with gamma radiation is more dangerous for external irradiation. However, it is less cumulative and disappears more rapidly from animals and soil, because it is more soluble and does not fix permanently in biological structures. Sr^{90} is a close analogue of calcium, and it can easily replace calcium in bones and soil. Calcium is part of the permanent structure of a body, and so Sr^{90} can remain in the body for many years, and in the soil for hundreds of years. This is why Sr^{90}, which has beta radiation, is considered the most dangerous product of nuclear bomb tests and of the nuclear industry.

If the nuclear disaster in the Urals really caused the contamination of hundreds or thousands of square miles of territory, this area must still be heavily contaminated by Sr^{90} and partly by Cs^{137} today. The soil, animals living in the soil, plants, insects, mammals, lakes, and fish, and all other forms of life in this area would still contain significant amounts of Sr^{90} and Cs^{137}, right up to today. The random distribution of radioactive isotopes during an accident of this type could cause the level of isotope concentration to vary enormously from place to place. In many areas the external and internal radiation could seriously threaten the life of many species, increasing the rate of mutations and mortality, and inducing many other changes. The extremely large contamination area would also create a unique biocoenosis, where genetic, population, botanical, zoological, and limnological research into the influence of radioactive contamination could be carried on in natural surroundings.

Critics of Tumerman's and my story can obviously ask: Why, then, did Soviet scientists miss this chance to study so many unique radiobiological and genetic problems, which this enormous radioactive environment (certainly the largest in the world) provided for long-term study?

The answer is very simple. Soviet scientists *did not miss this chance,* and more than a hundred works on the effect of Sr^{90} and Cs^{137} in natural plant and animal populations have been published since 1958. In most of these publications neither the cause nor the geographical location of the contaminated area is indicated. This is the unavoidable price of censorship. However, the specific composition of plants and animals, the climate, soil types, and many other indicators lead to the inevitable conclusion that the contamination is in the South Ural region. In one publication the Cheliabinsk region is actually mentioned (a slip-up in censorship). The length of observation (ten years in 1968, eleven in 1969, fourteen in 1971, etc.) reveals the approximate date of the original accident. Finally, the scale of the research (especially with mammals, birds, and fish) clearly indicates hundreds of square miles covered by rather heavy radioactive contamination in an area with several large lakes.

I cannot give here a comprehensive review of these radio-ecological and genetic works. But even a short summary of the most important publications clearly confirms the brief accounts of this environmental tragedy contained in my earlier article and in Lev Tumerman's report.

I have known about the nuclear waste explosion in the Urals area since 1958. My professor, Vsevolod Klechkovsky (now dead), was an expert in the use of radioactive isotopes and radiation in agricultural research. He was given the task at that time to set up an experimental station within the contaminated territory. The station would study the effect of radioactive isotopes on plant and animal life and monitor the so-called "secondary distribution" of contamination. Radioactive pollution of this type cannot be confined within the initial area—soil erosion and biological distribution constantly widen the radioactive region. The specific activity of contamination in the original area and in the new, neighboring ones declines, however, in time. I was approached by Klechkovsky with an offer of a job at this station, but I did not accept it, because of the secrecy involved. However, some junior research workers from the Department of Agrochemistry and Biochemistry of the K. A. Timiriazev Agricultural Academy, headed by Professor Klechkovsky, started to work at this station (and work there now in some senior positions).

All aspects of the work associated with this nuclear disaster from the very beginning were highly classified. There was no chance of publishing the results of any research carried out there. However, after Khrushchev's fall the situation changed slightly, because the whole tragedy could be considered the responsibility of Khrushchev's nuclear authorities. The chairman of the State Committee for Atomic Energy of the USSR, Professor Vasily Emelyanov, was dismissed from his post in 1965; some other high officials in both the civil and military branches of the atomic energy industry went as well. It was considered unnecessary, in 1965–1966, to acknowledge the occurrence of a catastrophe which had happened several years earlier. But at least the veil of secrecy that had surrounded the disaster was lifted. Many experts from the Soviet Academy of Sciences and other research establishments were permitted to start comprehensive research of the contaminated area and to publish their results in the Soviet academic journals.

This change of attitude also became possible because of the end of T. D. Lysenko's domination in biology and genetics. Several new research institutes and units on genetics, radiobiology, and ecology created in 1965–1966 applied heavy pressure to get access for research to this unique radioactive environment. These studies, unfortunately, were undertaken several years after the initial impact of the radioactive hazard on the biocoenosis, which included all levels of life—soil bacteria, soil algae, soil animals (worms, insects, and others), amphibians, reptiles, birds, rodents, large mammals (deer and others), water plants, different species of fish, perennial plants, trees, and so on. Farm plants and animals, as well as the human population, were also included in places of "secondary distribution," where the level of radioactivity was not so high as to make the evacuation of inhabitants obligatory.

I do not plan to cover the main results of all these works here. But I will indicate at least some of the published works, which give information about the approximate size of the area and about the contamination level of the environment, mainly by Sr^{90} and Cs^{137}.

Lakes

One of the first works, pointing to a possible serious industrial nuclear disaster and published by F. Rovinsky in 1965 in the Soviet journal *Atomnaya Energiya* (Atomic Energy) (vol. 18, 379–383), appeared to by purely mathematical. Its title, "The Method of Calculation of the Distribution of Radioactive Contamination in Water and Bottom Deposits of Non-Running-Water Lakes," was rather theoretical and the whole text was saturated with mathematical equations. This study was based on the measurements of radioactivity in two lakes which had been contaminated by industrial radioactive waste five years before the measurements began. (The paper was submitted for publication in May, 1964, which means that the work was completed about 1963.)

During the first months the isotope composition was complex, but later Sr^{90} was the only dominating isotope. The water radioactivity (the level in absolute figures was not indicated) declined quickly during the first two years because of absorption by the silt. But then some kind of equilibrium between the bottom silt deposits was established. The theoretical calculations and the experimental picture were almost identical. One can find hardly anything wrong with the whole work or the "experimental" contamination, except for the size of the two lakes referred to. "The experimental lakes," wrote the author, "were eutrophic type, the first was 11.3 square kilometers in area and the second 4.5

square kilometers, both almost round in shape" (p. 380). It is rather hard to believe that anybody of sound mind would contaminate two lakes of such size for the sake of confirming some mathematical calculations. However, I did not find any other research on these two particular lakes, except five years later, in the report of a study of "nest conservatism" among some species of water birds.

The third contaminated lake appeared in two papers by A. I. Ilenko published in *Voprosy Ikhtiologii* (Problems of Ichthyology) (vol. 10, [1970], 1127–1128, and vol. 12, [1972], 174–178). The author studied the distribution of Cs^{137} and Sr^{90} in water, plankton, water plants, and different fish species. The research was carried out during 1968–1970, but the lake had been contaminated many years before. The concentration of both isotopes in the lake mentioned in these papers varied every month, depending greatly on seasonal conditions (with maximums during October and July). These variations could be typical only for a running-water lake with a contaminated basin. During the summer of 1969 the concentration of Sr^{90} in the water was 0.2 microcurie per liter and the concentration of Cs^{137} 0.025 microcurie per liter. Both figures were a *hundred times* higher than contamination levels in some small experimental ponds made especially for research purposes in certain earlier studies in the USSR and other countries. The purpose of Ilenko's work was to study the food chains among different forms of life in the lake. To make this possible the population balance in the lake should not be seriously affected. The largest and final link in the food chain was pike (*Esox lucius*). To measure the isotope concentration in their bones and muscles, more than one hundred pike had been studied, some very large (twenty-five to thirty pounds each). The lake was not "rich," because only four species of fish were found there. For food-chain research, the whole population of pike had to be at least ten to twenty times larger, and the size of a lake with this number of large pike must be between 10 and 20 square kilometers. To contaminate such a lake by Sr^{90} up to 0.2 microcurie per liter (if it be non-running water and not deep), one would need at least 50,000 curies, an amount far too large for experimental purposes. For a running-water type of lake the amount would have to be many times more. However, Ilenko calculated that in the water plants, plankton, and silt the total amount of Cs^{137} and Sr^{90} was more than *a thousand times higher than in the water,* and the concentration of Cs^{137} in the water plants, for example, varied from 10 to 38 microcuries per kilogram.* The concentration of Cs^{137} in pike muscles also was much higher (from 100 to 1,500 times in different months) than in the water. Thus the total minimal amount of Sr^{90} and Cs^{137} in the whole lake would be on the order of millions of curies (one curie is a unit of radioactivity defined as that amount of any radioactive material that has the same disintegration rate as one gram of radium 226, i.e., 3.7×10^{10} disintegrations per second), and this enormous amount of radioactivity filtered into the lake from the lake's basin! It is, however, well known that the soil fixes strontium very strongly and that only a small fraction could be filtered with the soil (about 5 to 6 percent over several years). It is, of course, impossible to know precisely how many hundreds of millions of curies of Sr^{90} and Cs^{137} had to be fixed in the lake's basin to make it possible to accumulate such an enormous amount of radioactivity in a running-water lake. There were no precedents for such research. Could anyone imagine that this amount of radioactive material would be distributed over the area surrounding the lake just for "experimental" purposes?

---

* The lakes in the Ural area usually have very thick bottom silt deposits. In Rovinsky's work, mentioned earlier, the total amount of Sr^{90} in the bottom silt was at least ten times higher than in the water when equilibrium was reached. However, in his work the lakes were non-running-water types. In the case studied by Ilenko, the lake had an intensive turnover of water supply (the Sr^{90} concentration could decrease or increase more than 400 percent within one month). These conditions make bottom silt and the water plants the main accumulators of radioactive materials. And this accumulation started many years before the beginning of Ilenko's experiments.

Mammals

There are many published works dealing with different species of mammals living in the contaminated area (the levels of contamination were usually the same in different experiments from 0.2 to 1.0, from 0.1 to 1.5, and from 1.8 to 3.4 millicuries of Sr^{90}, and 4 to 7 microcuries of Cs^{137} per square meter in 1965–1969). A. I. Ilenko and his collaborators also carried out several research projects with mammals at the same time (1968–1970) they were doing the work on the lake's population.[*] In two studies of mammals, where food chains were also the main research aim (research of this type must be carried out without any serious change in the population balance), about 2,000 individual animals of fifteen different species were killed. Small animals (mice, rats, rabbits) are poor indicators of the size of a research area. However, these two papers—*Zoologicheskii Zhurnal* (Journal of Zoology), vol. 49 (1970), 1370–1376; *Zhurnal Obshchei Biologii* (Journal of General Biology), vol. 31 (1970), 698–709—reported the shooting of twenty-one deer which also were living within the contaminated environment. This end of the food chain is indeed rather revealing, because the shooting of deer should be done without any serious depletion of their natural population and species ratios. One can suggest that at least a hundred were available. Deer normally migrate over great distances, especially during winter. The average territory for such work should not be less than about 100 square miles. The level of soil contamination by Sr^{90}, between 1.8 and 3.4 millicuries per square meter, is also much higher than any possible "experimental" contamination.

[*] Since samples of fish and animals were taken continuously, the whole research certainly was carried out in the same environment. These figures on radioactivity are given for work with plants (at least six to seven different groups of plant communities, including forests), soil animals, etc. Most experimenters refer to Ilenko's work, which indicates that the same region is being studied. Explanations of the cause of the contamination either vary or are not given.

About one million curies of Sr^{90} are necessary to obtain such an "experimental" field. Works by other authors, in which plants, soil, and soil animals were studied (identical figures for the radio-activity and references to Ilenko's works indicate the same local-ity for the "experimental" area) also prove an area of geographical scale, not just a fenced field. Most of the plants were of the cross-pollinator type. The contaminated territory had many different soil types, varying from gray forest type to black soil. It consisted of meadows, hills, plains, and various kinds of forests. In general, within the same contamination area there were at least seven or eight different biocoenoses.

Population Genetics and Radiation Genetics

The research on population genetics (bacterial, plant, and animal) was carried out by a large team of research workers, headed by academician N. P. Dubinin. The main results have been published in fourteen or fifteen papers since 1968 in *Genetika* (Moscow) and in other journals. Our task of giving at least a short summary of this research is simplified by a review of the results, published in Russia's equivalent of the *Annual Review on Genetics* in 1972. (N. P. Dubinin and others, *Uspekhi Sovremennoi Genetiki* [Moscow], vol. 4 [1972], 170–204).

The contaminated area where the various research studies had been carried out was the same. The authors acknowledged that the level of radioactivity was 1.8 to 3.4 and 1 to 1.5 millicuries per square meter, with a reference to Ilenko's works. They also ac-knowledged that the contamination had not been specially ar-ranged for their experiments and that they were able to start their radiobiological and genetical observations (the frequency and pattern of chromosomal aberrations, comparative radiosensi-tivity, selection of radioresistant forms, and others) *only five to seven years after the organisms, selected for research purposes, had already lived in the radioactive environment*. This was a def-inite disadvantage for population genetics work, because the earlier stages of adaptation had been missed and the initial level of irradiation by the mixture of short-lived and long-lived isotopes was unknown. In spite of these methodological aberrations, the authors, in all their studies, were able to find a selection of more resistant forms and some other genetic population changes in soil algae (Chlorella), many different plants (mostly perennial), and rodents (different species of mice).

The special aspect of the work, which I want to emphasize, is the size of the research area. In their work on rodents (*Clethrionomus rutilus,* usually found in meadows, and *Adodemus sylvaticus,* living in forests) the authors acknowledged that they started their genetic research with a population that had *already lived thirty generations in a radioactive environment.* For genetic population work one has to be certain that the individual animals being examined are the true ancestors of these animals, and lived in the radioactive area when the original radioactive contamination occurred. Rodents do not migrate very far during their adult life. However, every new generation can migrate from the parental environment over ever greater distances, usually within 1,000 meters or more. During thirty generations, migration could be as far as 20 to 30 kilometers, which means 400 to 900 square kilometers of radioactive environment. The authors of these works did not give the exact size of their research area, but they acknowledged that all animals under study really lived in the radioactive zone over all these years. And in the table where the figures for the frequency of aberrations of chromosomes and the concentration of Sr^{90} in bones are shown, the level of contamination of the environment is given as 1,800–3,500 curies per square kilometer for one experimental group and 1,000–1,500 curies per square kilometer for another group of animals.

For different trees this level of radioactivity had been too high and some species did not survive. However, for single-cell soil algae (Chlorella), which are very resistant, radioactivity could reach a very high level before there was any genetic damage.

As I have already indicated, the authors did not contaminate the area for the purpose of their research, and started work several years after it had taken place. The first samples of Chlorella (*Ch. vulgaris*) were taken after five years of radioactive contamination, which meant that there had been more than two hundred generations of this algae. This work was done in a different place, because the radioactivity of the soil was much higher and is indicated not in square meters but in *kilograms* of soil. There were variations in soil activity, but the maximum radioactivity was 1.10^{10} disintegrations per kilogram of soil per minute, which comes to about 50 millicuries. If this activity is calculated by the square meter, it is about 5 curies for a surface layer of 8 to 10 centimeters!

The contaminated area in which all this work was done had a very uneven distribution of radioactivity. The work on plants and animals was carried out in places where these animals and plants could live many generations. Places where they could not survive were certainly not explored so well. But the fact of the existence of such areas in the same location has been acknowledged in academician N. Dubinin's autobiography.* Stating briefly (p. 330) that his group did long-term research in an area "contaminated by high doses of radioactive substances," he wrote: "Part of the species died in this environment, part suffered and declined slowly, but some evolved towards a higher resistance."

In all these works the nature of the species of microbes, plants, insects, and animals (more than two hundred species are referred to in all) could easily indicate the approximate geographical location. The mixture of European and Siberian species could easily point to the Urals zone. I had intended to check this in the multivolume editions of *Plants of the USSR* and *Animals of the USSR*, but this detective work was obviated by the accidental acknowledgment in one of the recent works of Ilenko and his col-

* See p. 226, above.

laborators that the animals for their work had been collected in the Cheliabinsk region (*Radiobiologiya,* vol. 14 (1974), 572–575). This particular work had been carried out during the fall of 1971 and the animals had lived in the radioactive environment for fourteen years, hence since the autumn of 1957.

All these publications represent rather a small part of the research data received from this contaminated environment and published in different Soviet scientific journals. British or American nuclear authorities probably put more trust in the information they receive from monitoring global fallout and by space satellite surveillance. They certainly do not read such Russian journals as *Voprosy Ikhtiologii, Genetika,* or *Zoologicheskii Zhurnal.*

There are probably very few foreign scientists, as well, who read these journals regularly, and even fewer who can understand the meaning of many methodological omissions. This is why so many experts were puzzled and doubtful about my article in *New Scientist* last November. Science fiction or not, many millions of curies of Sr^{90}, Cs^{137}, and other radioactive isotopes did contaminate a very large area in the South Urals region where the first Soviet military nuclear reactors were built between 1947 and 1949. The nature of the contamination certainly excludes a reactor accident or real atomic explosion. The available facts in the published materials are much more consistent with an accident of a nuclear waste disposal site. How it happened and what was the real human price of this accident have not yet been published. Soviet secrets can often be very long-lived.*

*The nuclear disaster in Cheliabinsk region during the winter of 1957–1958 has recently been confirmed by CIA documents released under the Freedom of Information Act. Articles and comments about these documents had been published in the *Washington Post,* the *New York Times,* and other newspapers on November 26, 1977.

Bibliography and Notes

(1) Komkov, G. D., B. V. Levshin, L. K. Semenov. *Akademia Nauk SSSR. Kratkii Istoricheskii Ocherk* (Academy of Sciences of the USSR. Short History). Moscow: Nauka, 1974.

(2) Crowther, J. G. *Soviet Science.* Harmondsworth, Eng.: Penguin, 1942.

(3) Ashby, Eric. *Scientist in Russia.* Harmondsworth, Eng.: Penguin, 1947.

(4) Vacinich, Alexander. *The Soviet Academy of Sciences.* Stanford University Press, 1956.

(5) Joravsky, D. *Soviet Marxism and Natural Science, 1917–1932.* New York: Columbia University Press, 1961.

(6) Graham, Loren R. *Science and Philosophy in the Soviet Union.* New York: Knopf, 1972.

(7) Parry, Albert. *The Russian Scientist.* New York: Macmillan, 1973.

(8) Medvedev, Zhores A. *The Rise and Fall of T. D. Lysenko.* New York: Columbia University Press, 1969.

(9) Medvedev, Zhores A. *The Medvedev Papers.* London: Macmillan; New York: St. Martin's, 1971.

(Part I is available as a paperback under the title *National Frontiers and International Scientific Co-operation* [Nottingham, Eng.: Spokesman Books, 1975]. Part II is also available as a paperback edition under the title *The Secrecy of Correspondence Is Guaranteed by Law* [Nottingham: Spokesman, 1975].)

(10) See *Communist Party of the USSR in Statements and Decrees of Congresses, Conferences and Plenums of Central Committee*, vol. 2, p. 52. Moscow, 1970 (in Russian).

(11) Reznik, S. *Nikolai Vavilov: A Biography*. Moscow: Molodaia Gvardiia, 1968.

(12) *Ustavy Academii Nauk SSSR: 1924–1974*. Moscow: Nauka, 1975.

(13) *Deciat' Let Sovetskoi Nauki, 1917–1927* (Ten Years of Soviet Science). Moscow and Leningrad, 1972.

(14) Ioffe, A. F. "Russkaia Nauka za Granizei" (Soviet Science Abroad), *Pravda*, July 21, 1924.

(15) Bliacher, L. Ya., ed. *Istoria Biologii* (History of Biology). Moscow: Nauka, 1975.

(16) Lerner, I. M. and W. J. Libby. *Heredity, Evolution, and Society*, 2nd ed. San Francisco: Freeman, 1976.

(17) Koltzov, N. K. *Organizatsiia Kletki* (The Organization of the Cell) (Collected Works). Moscow: Biomedgiz, 1935.

(18) Medvedev, Roy A. *Let History Judge*. New York: Knopf, 1971; London: Macmillan, 1972.

(19) Stalin, I. V. *Sochineniia* (Collected Works), vol. 12, p. 14. Moscow: Gospolitizdat.

(20) Graham, L. R. *The Soviet Academy of Sciences and the Communist Party, 1927–1932*. Princeton University Press, 1967.

(21) Ioffe, A. F. *Mezdunarodnye Sviazi Sovetskoi Nauki, Tekhniki i Kultury* (International Connections of Soviet Science, Technology and Culture) (1917–1932). Moscow: Nauka, 1975.

(22) Conquest, Robert. *The Great Terror*. New York and London: Macmillan, 1968; 2nd ed., 1973.

(23) Carr, E. H. *A History of Soviet Russia*. 7 vols. New York: Macmillan, 1951–1960.

(24) Joravsky, David. *The Lysenko Affair*. Cambridge, Mass.: Harvard University Press, 1970.

(25) The detailed story of A. N. Tupolev's arrest and engineering work in a special prison before the war and during the war was described in a short, fascinating *samizdat* work, "Tupolevskaya Sharaga" (*sharaga* is a slang name for prison-research establishments). This work was signed by the pseudonym "A. Sharagin." It was published in 1971 in Russian by the émigré publishing house Possev, in

Frankfurt am Main. However, the real name of the author was established and acknowledged later: Professor Georgy Oserov.

(26) Gasarian, S. "Eto Ne Dolzhno Povtoritsa" (This Must Not Be Repeated). Manuscript in Russian. This work has been circulating in *samizdat* since 1964. Roy Medvedev in his *Let History Judge* (18) often quotes from Gasarian's book.

(27) Medvedev, Roy A. "O Staline i Stalinisme" (About Stalin and Stalinism). Manuscript, 1977.
(To be published by Oxford University Press approximately in 1979.)

(28) Golovin, I.N. *I. V. Kurchatov*. Moscow: Atomizdat, 1967.

(29) Astashenkov, P. *Kurchatov*. Moscow: Molodaia Gvardiia, 1967.

(30) Medvedev, Zhores A. "A New Method of Autoradiography for Investigation of the Localization of Synthesis of Proteins and Nucleic Acids in Plants," in *Proceedings of the First International Congress on Radioisotopes in Scientific Research* (UNESCO), vol. 3, 648–666. London and New York: Pergamon Press, 1958.

(31) Sakharov, Andrei. *Sakharov Speaks*. New York: Knopf, 1974.

(32) Dorozynskii, A. *The Man They Wouldn't Let Die*. London: Secker & Warburg, 1967.

(33) Smolders, Peter. *Soviets in Space*. Guildford, Eng.: Lutterworth Press, 1973.

(34) Lepeshinskaya, Olga. *Proischozhdenie Kletok iz Zhivogo Veshestva i Rol Zhivogo Veshestva v Organisme* (Origin of Cells from Living Substance and the Role of Living Substance in the Organism). Moscow, 1945.
(The first edition of this book was published with an introduction by T. D. Lysenko. Later several other editions were published and the book was awarded the Stalin Prize.)

(35) Boshian, G. M. *O Prirode Virusov i Mikrobov* (About the Nature of Viruses and Microbes). Moscow: Selchozgis, 1949; 2nd ed., 1950.

(36) Aslan, Anna. "A New Method for Prophylaxis and Treatment of Aging with Novocain: Eutrophic and Rejuvenating Effects," *Therapiewoche*, vol. 7 (1956), 14–22.

(37) *Narodnoie Khoziaistvo SSSR v 1974* (Economy of the USSR in 1974), annual statistical report. Moscow: Statistika, 1975.

(38) Medvedev, Roy A., and Zhores A. Medvedev. *Khrushchev: The Years in Power*. New York: Columbia University Press, 1976; London: Oxford University Press, 1977.

(39) Dobrov, G. M. *Nauka o Nauke* (The Science of Science). Kiev: Naukova Dumka, 1970.

(40) Medvedev, Zhores A. "New President of Soviet Academy," *Nature* (London), vol. 258 (1975), 566. (A reply to the article by Vera Rich published in *Nature,* vol. 258 (1975), 377.

(41) Khrushchev, Nikita S. *Khrushchev Remembers: The Last Testament.* Translated and edited by Strobe Talbott. Boston: Little, Brown, 1974.

(42) Medvedev, Zhores A. "Two Decades of Dissidence," *New Scientist,* vol. 72 (1976), 264–267.

(43) Patterson, Walter C. *Nuclear Power.* Harmondsworth, Eng.: Penguin, 1976.

(44) Zaleski, E., J. P. Kozlowski, H. Wienert, R. W. Davies, M. J. Berry, and R. T. Amann. *Science Policy in the USSR.* Paris: Organization for Economic Cooperation and Development, 1969.

(45) *The State of Soviet Science.* Edited by the editors of *Survey: A Journal of Soviet and East European Studies.* Cambridge, Mass.: M.I.T. Press, 1965.

(46) Hutchings, Raymond. *Soviet Science: Technology, Design, Interaction, and Convergence.* London: Oxford University Press, 1976.

(47) Roskin, G. "Toxin Therapy of Experimental Cancer: The Influence of Protozoan Infection upon Transplanted Cancer," *Cancer Research,* vol. 6 (1946), 363–365.

(48) Zhebrack, A. R. "Soviet Biology," *Science,* vol. 102 (1945), 357–358.

(49) Sakharov, A. *About My Country and the World.* New York: Knopf, 1976.

(50) Solzhenitsyn, A. *Amerikanskie Rechi* (American Speeches). New York: YMCA Press, 1975.
    (Also published in Russian in a special Russian edition of *News of Free Trade Unions* [AFL-CIO], vol. 30, nos. 7–8 [1975]. Several interviews with Solzhenitsyn made for the BBC in England have been published under the title *Warning to the Western World* [London: Bodley Head, 1976].)

(51) Medvedev, Roy A. *Political Essays.* Nottingham, Eng.: Spokesman Books, 1976.

(52) Coates, Ken, ed. *Détente and Socialist Democracy: A Discussion with Roy Medvedev.* Nottingham, Eng.: Spokesman Books, 1975.

(53) Urban, G. R., ed. *Détente.* New York: Universe Books, 1976.

(54) *Détente: Hearings before the Subcommittee on Europe of the Committee on Foreign Affairs, House of Representatives, 93rd Congress.* Washington: U.S. Government Printing Office, 1974.

(55) *Détente: Hearings before the Committee on Foreign Relations,*

United States Senate, 93rd Congress. Washington: U.S. Government Printing Office, 1974.

(56) Medvedev, Zhores A. In *Détente: Hearings before the Senate Committee on Foreign Relations,* pp. 413–454.

(57) Nichols, R. "Career Decision of Very Able Students," *Science,* vol. 144 (1964), 1315–1319.

(58) *Narodnoie Khoziaistvo SSSR v 1975,* annual statistical report. Moscow: Statistika, 1976.

(59) Zhukov, G. K. *Vospominaniia i Razmyshleniia* (Memoirs and Thoughts). Moscow: Voenizdat, 1969.

(60) Shtorch, D. "The Chance Which America Offers," *Time and We* (*Vremya i my*), no. 1 (1975), 145–164 (in Russian). Tel Aviv, Israel.

(61) Murray, B., and M. E. Davies. "Détente in Space," *Science,* vol. 192 (1976), 1067–1074.

(62) Solzhenitsyn, A. *Amerikanskie Rechi* (50), p. 15. Quotation from a speech at a meeting at the Hilton Hotel, Washington, D.C., on June 30, 1975.

(63) Etkind, E. *Dissident Ponevole* (Dissident Against One's Will), as quoted in *Kontinent,* no. 7 (1975). Germany.

(64) See *Izvestia,* no. 17 (18165), Jan. 23, 1976.

(65) See *Bulletin Vysshei Attestazionnoi Comissii SSSR* (Bulletin of the Highest Qualification Commission of the USSR), no. 2 (1976).

(66) *SSSR v Zifrach in 1976* (USSR in Figures in 1976). Moscow: Statistika, 1977.

(67) Edsall, John T., ed. *Scientific Freedom and Responsibility.* Report of the AAAS Committee on Scientific Freedom and Responsibility. Washington: American Association for the Advancement of Science, 1975.

(68) Byrnes, Robert F. *Soviet-American Academic Exchanges, 1958–1975.* Bloomington: Indiana University Press, 1976.

(69) *Nature* (London), vol. 262 (1976), 535–536.

(70) *Science,* vol. 194 (1976), 166.

(71) *Science,* vol. 194, no. 4267 (1976).

(72) Wilczynski, J. *Technology in Comecon: Acceleration of Technological Progress through Economic Planning and the Market.* London: Macmillan, 1974.

Index